MAGNETIC AND ELECTROMAGNETIC SHIELDING

MAGNETIC AND ELECTROMAGNETIC SHIELDING

T. RIKITAKE

Dept. of Earth Sciences, Nihon University, Tokyo, Japan

Terra Scientific Publishing Company / Tokyo

D. Reidel Publishing Company

A MEMBER OF THE KLUWER ACADEMIC PUBLISHERS GROUP

Dordrecht / Boston / Lancaster / Tokyo

Library of Congress Cataloging-in-Publication Data

CIP

Rikitake, Tsuneji, 1921–
Magnetic and electromagnetic shielding.

Bibliography: p.
Includes index.
1. Shielding (Electricity) 2. Magnetic shielding.
I. Title.
TK454.4.M33R58 1987 621.34 86-31383
ISBN 978-90-481-8433-0

Published by Terra Scientific Publishing Company (TERRAPUB),
307 Shibuyadai-haim, 4-17 Sakuragaoka-cho, Shibuya-ku, Tokyo 150, Japan,
in co-publication with D. Reidel Publishing Company, Dordrecht, Holland

Sold and distributed in the U.S.A. and Canada
by Kluwer Boston Inc.,
190 Old Derby Street, Hingham, MA 02043, U.S.A.,
in Japan by Terra Scientific Publishing Company (TERRAPUB),
307 Shibuyadai-haim, 4-17 Sakuragaoka-cho, Shibuya-ku, Tokyo 150, Japan

In all other countries, sold and distributed
by Kluwer Academic Publishers Group,
P. O. Box 322, 3300 AH Dordrecht, Holland

D. Reidel Publishing Company is a member of the Kluwer Academic Publishers Group

Printed in Japan

PREFACE

When the author was in the navy some forty years ago, he was involved in designing magnetic devices and in installing them in ships. He found that at the bridge of a ship the geomagnetic field is very much weakened and disturbed. Inside a submarine, the field is shielded so strongly that a magnetic compass is unusable.

It was sometimes required to place magnetic equipment very close to an electric line which powered a submarine and through which a d.c. electric current of several hundred amperes flowed. Because of the strong magnetic field arising from such a current, watches were magnetized so much that they did not function and a magnetic compass could not be used.

These difficulties led the author to investigate magnetic shielding effects in general. Unfortunately, however, the construction of a general theory of magnetic shielding was not accomplished because of mathematical difficulties. Only the theory of magnetic shielding relating to spherical and cylindrical shells of high magnetic permeability could be formulated at that time. Even the shielding effect due to multiple spherical shells was difficult to determine since it required the solution of a complicated set of simultaneous equations.

Under the circumstances, the author was forced to keep these unsolved problems of magnetic shielding in mind all through his academic life after the war hoping to tackle them some day.

The author has been interested in studying the electrical state in the earth's interior at the Earthquake Research Institute, University of Tokyo, where he spent more than thirty years, after being demobilized from the navy. As such studies rely largely on electromagnetic induction within the earth by time-dependent magnetic fields, the study of electromagnetic shielding has naturally been one of his concerns.

Although the interest is mostly geophysical, the development in such a study can readily be modified so that it is applicable to practical problems of electromagnetic shielding. For instance, shielding against hum noise arising from commercial electric power lines can be evaluated by applying theory developed for geophysical problems.

When the author moved to the Department of Applied Physics, Tokyo Institute of Technology in 1975, he thought that a useful way to complete the theory of magnetic and electromagnetic shielding was to ask students in the department to work on particular problems of shielding, one by one. Their work is organized in such a fashion that the summary of their theses forms an approximately complete theory of shielding.

Fortunately, the use of high-speed computers has now become common place, so that problems which the author failed to solve a few tens of years ago have become tractable even by undergraduate students. Of considerable benefit in the solution of practical problems is the application of the relaxation method for solving partial differential equations under proper boundary conditions. Using the method, problems of shielding by rectangular and octagonal shells can be solved. Even shells with a hole can be treated, as can shielding by a cylindrical shell of finite length.

The author now believes that much of the theory of magnetic and electromagnetic shielding has at last been completed. The contents of this book may be of some use when considering various problems of magnetic and electromagnetic shielding.

F. Sasada, K. Tsukuda, M. Suzuki, S. Morimoto and K. Watanabe, former students of the author, worked on some of the material in this book. Their effort is appreciated. Professor Y. Honkura and Dr. H. Tanaka helped these students in accomplishing their work. The author is grateful to them also. N. Ohshiman and Y. Fujimori computed some of the solutions to electromagnetic induction in single- and two-layer models.

Mr. K. Oshida at the Terra Scientific Publishing Company helped the author in the course of writing the manuscript. The author thanks him for his help. The Japanese Ministry of Education defrayed part of the expenses for publishing this book. The author thanks the Ministry for the financial aid.

CONTENTS

Chapter 1

INTRODUCTORY REMARKS

Electric, magnetic and electromagnetic shielding often plays an important role when making electric and magnetic measurements of high sensitivity. For example, a rock sample should be stored in a very weak magnetic field in order to avoid possible contamination by an external magnetic field. In studies of rock magnetism, it is customary to surround samples by shells or sheets of high magnetic permeability. In order to calibrate a highly sensitive magnetometer, it is often necessary to place it in a fairly large shielding room in which the geomagnetic field and its fluctuations are very much reduced.

In the case of rocket- or satellite-borne magnetic instruments, it is vitally important to protect them from magnetic fields arising from components such as electric motors and batteries which may be mounted not far from the instruments concerned. This is also the case for ocean bottom magnetometers which are currently in widespread use. Further, it is well known that for accurate time-keeping a watch should not be placed in a strong magnetic field.

In addition to the necessity of shielding from static magnetic fields, it is also highly important for precise magnetic and electronic instruments to be shielded from external oscillating magnetic fields. For instance, the hum noise from commercial electric power lines affects not only scientific instruments but also domestic audio-equipment. It is sometimes no easy matter for high-gain electronics to eliminate hum noise from nearby electric transformers.

Turning to naturally fluctuating magnetic fields, it can be seen that electromagnetic induction in the ionosphere, the oceans and the earth's interior, all of which are to some extent electrically conducting, has an important bearing on geomagnetic and geoelectric studies. For example, an oscillating magnetic field having a period of 1 min. or shorter, arising from outside the earth, seems likely to be shielded almost completely by a uniformly conducting shell equivalent to a worldwide ocean of uniform depth 1000 m. A 100-year period geomagnetic secular variation produced by some time-dependent electric current originating within the earth's core (that part of the earth below a depth of 2900 km), seems likely to be weakened to some extent as it diffuses through the earth's mantle to the surface.

On the other hand, electromagnetic induction studies provide a powerful means of inferring electrical properties within the earth, the moon and possibly other terrestrial planets. The distribution of electrical conductivity determined by such studies aids an understanding of the physical state within these celestial bodies.

Noting the above varied examples, it would seem useful to construct an overall theory of magnetic and electromagnetic shielding for the purpose of applying it to practical electronic technology as well as to geophysical problems. This book is written in the hope of fulfiling such a requirement. It should be pointed out, however, that what will be developed has little to do with practical shielding or screening techniques widely discussed in electrical engineering; to give but two examples, the transmission of electricity and the mutual induction between two transmission lines.

What the author intends to present in this book are the basic facts which can be deduced by solving the fundamental equations of electricity and magnetism. The author believes that no such detailed presentation exists elsewhere in the current literature. It will be beyond the scope of this book and the author's capability to deal with highly technical matters in electrical engineering.

In Chapter 2, the shielding of a static magnetic field by spherical, cylindrical and rectangular shells having high magnetic permeability will be discussed. It will be shown that efficient shielding cannot be achieved simply by increasing the shell thickness. Rather, it is much more efficient to make use of multiple shells, there being an optimal configuration achieving the most efficient shielding. No discussion of this kind has been presented in the existing literature as far as the author knows, so that this chapter should be of use when considering practical designs for shielding static magnetic fields.

The shielding effect of spherical and cylindrical shells to the magnetic field arising from inside the shells will also be studied. It is remarkable that the shielding effect for an internal magnetic dipole is exactly the same as that for an external uniform magnetic field.

The author is proud to present in this chapter the shielding effect of cylindrical shells having a finite length. Although the shielding effect of an infinitely long cylindrical shell has been presented in existing text books, it is important to determine the effect of the ends of a cylindrical shell for applying the theory to practical problems. A study of such problems solved by the relaxation method will be presented for cases of axial symmetry. It will be shown how long a cylindrical shell must be in order to achieve a shielding effect equivalent to that for an infinitely long shell. The study will be made for two cases: when the inducing field is parallel and when it is perpendicular to the cylinder axis.

It will be of interest to examine the shielding effect of a rectangular shell to an external magnetic field. It will be shown in this chapter that the leakage of magnetic lines of force is considerable at the corners of the shell. In order to achieve high-quality shielding, therefore, it is necessary to use a shell which has a smooth shape. Rectangular shells with a hole will also be studied. The conclusion obtained plays an important role in designing a system for shielding which must necessarily have some holes for introducing lead wires and other devices. To the author's knowledge, no other literature discussing the effect of such a hole in a magnetic shell is available. It is remarkable that all these results for magnetic shielding by complete and incomplete rectangular shells can be obtained by solving problems using the method of relaxation.

The discussion in this chapter can be applied to problems of shielding to a static electric field by replacing magnetic field by electric field, magnetic potential by electric potential, magnetic lines of force by electric lines of force and magnetic permeability by dielectric constant.

The shielding effect of an electrical conductor, either magnetically permeable or non-permeable, will be studied in Chapter 3 when the inducing magnetic field is time-dependent. It is interesting to note that a conductor tends to deflect magnetic lines of force from it, whilst highly-permeable magnetic matter tends to attract magnetic lines of force. The former effect is useful for rarefaction of magnetic lines of force. On the other hand, the latter effect leads to concentration of magnetic lines of force in the magnetic material, but the magnetic flux density immediately outside is reduced. In order to accomplish efficient shielding for an oscillating magnetic field, both effects should be properly combined.

In a similar manner to the last chapter, shielding by plane, spherical, cylindrical, square and octagonal shells for an oscillating magnetic field will be presented in Chapter 3. It is the author's intention to make this chapter useful for designing a system of electromagnetic shielding.

One of the sections of this chapter will be devoted to the theory of electromagnetic induction in, and so electromagnetic shielding by, thin shells and sheets of various configuration. There are problems of electromagnetic induction for which a conductor can be treated as an extremely thin shell or sheet, namely when the thickness of the conductor in question is very much smaller than the wavelength of the inducing magnetic field. The theory of electromagnetic induction then becomes very much simplified. The solutions advanced for thin shells and sheets will be applied to the shielding effect of a highly conducting plate, an ocean surrounding the earth and other examples.

Discussion concerning electromagnetic induction in two-layer conductors completes this chapter. Although this does not apply to practical shielding devices, the presentation will be of some interest for electromagnetic studies of the earth.

In Chapter 4 a general theory of electromagnetic induction in a non-uniform shell or sheet will be presented. Although no general theory of electromagnetic induction in a conductor having a heterogeneous conductivity can be formulated, because we cannot treat electric charge distribution and magnetic field separately, it is possible to put forward a general theory in the case of a thin shell or sheet. This chapter has no application to practical shielding. However, attention will be drawn to a rather unusual resonance-like phenomenon of the electromagnetic response of a conducting shell or sheet having a particular distribution of conductivity.

In Appendixes, the author will present a few particular problems of magnetic and electromagnetic shielding which are logical extensions of the discussions in the foregoing chapters. To preserve clarity in the text, they are presented separately in the Appendixes.

Electromagnetic induction problems that will be presented in this book are limited to cases in which electric displacement currents can be ignored, so that the shielding effect for rapid magnetic changes such as radio waves will not be studied. As the skin depth of the induced electric currents is small in such cases, it is the author's feeling that no sophisticated devices would be needed for shielding rapidly changing electric or magnetic fields.

Chapter 2

MAGNETIC SHIELDING

The aim of this chapter is to examine features of static magnetic fields in the presence of magnetic matter of high permeability. Emphasis is naturally laid on the shielding effect arising from rearrangement of magnetic lines of force.

In addition to the shielding effect due to a spherical shell and a cylindrical shell of infinite length, as often found in text books on magnetism, it is planned in this chapter to determine the effect due to a cylindrical shell of finite length and an infinitely long cylinder with square cross section. An interesting feature of magnetic lines of force near the corner of a shell of finite length will also be studied. The new shielding results stem mainly from work conducted by the author and his group at the Tokyo Institute of Technology and are not published in other text books.

2.1 Fundamental Equations and Boundary Conditions

According to the fundamental equations of electricity and magnetism, static magnetic field $\boldsymbol{H}$ and magnetic flux density $\boldsymbol{B}$ satisfy

$$\mathrm{curl}\,\boldsymbol{H}=0 \tag{2.1}$$

and

$$\mathrm{div}\,\boldsymbol{B}=0\,, \tag{2.2}$$

respectively. $\boldsymbol{B}$ is related to $\boldsymbol{H}$ by

$$\boldsymbol{B}=\mu\boldsymbol{H} \tag{2.3}$$

where μ is magnetic permeability.

Equation (2.1) makes it possible to put

$$\boldsymbol{H} = -\mathrm{grad}\ W \tag{2.4}$$

where W is called the magnetic potential. We therefore have from (2.2) and (2.3)

$$\mathrm{div}\ (\mu\ \mathrm{grad}\ W) = 0. \tag{2.5}$$

When μ is constant, (2.5) reduces to

$$\nabla^2 W = 0. \tag{2.6}$$

In addition, at a boundary between two media of different permeabilities the normal component of $\boldsymbol{B}$ and the tangential component of $\boldsymbol{H}$ must be continuous.

In order to determine the distribution of magnetic field in a space containing ferromagnetic substances, it is required to solve Eq. (2.6) under the above boundary conditions, along with an appropriate condition at a boundary surrounding the space considered.

2.2 Magnetic Shielding by Spherical Shells

2.2.1 Single shell

2.2.1.1 Shielding ratio for an external magnetic field

To obtain the distribution of the magnetic field in the presence of a uniform shell is a classical problem in magnetic field theory. Although the solution can readily be found in most text books, we first deal with a single shell because the study illustrates the basic method which may be generalized to cases of multiple shells.

Consider a uniform shell of which the outer and inner radii are respectively denoted by R_1 and R_2 as shown in Fig. 2.1. Spherical polar coordinates R, θ and ϕ are adopted here. The solutions of Eq. (2.6) in regions 0 ($R > R_1$), 1 ($R_1 > R > R_2$) and 2 ($R_2 > R$) are respectively given as

$$\begin{aligned} W_0 &= \left(-H_0 R + \frac{I_0}{R^2}\right)\cos\theta && R > R_1 \\ W_1 &= \left(-H_1 R + \frac{I_1}{R^2}\right)\cos\theta && R_1 > R > R_2 \\ W_2 &= -H_2 R\cos\theta && R_2 > R \end{aligned} \tag{2.7}$$

where I_0, H_1, I_1 and H_2 are to be determined by the boundary conditions

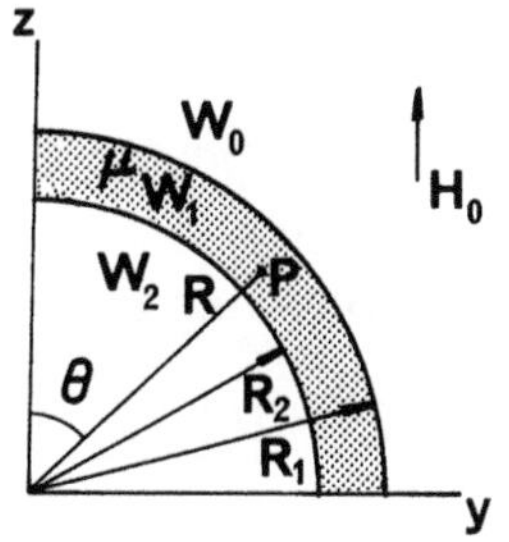

FIG. 2.1 A spherical shell of magnetic permeability μ placed in a uniform magnetic field H_0. W_0, W_1 and W_2 are the magnetic potentials defined for regions $R_1<R$, $R_2<R<R_1$ and $R<R_2$, respectively.

in terms of H_0. No term proportional to R^{-2} is included in the third equation because the potential should be finite at $R=0$. It is assumed here that there is a magnetic field H_0 parallel to the $\theta=0$ axis at infinity.

When the magnetic permeabilities are taken as μ in the shell and 1 in free space, the boundary conditions become

$$\begin{aligned} &W_0=W_1, \quad \frac{\partial W_0}{\partial R}=\mu\frac{\partial W_1}{\partial R} && R=R_1 \\ &W_1=W_2, \quad \mu\frac{\partial W_1}{\partial R}=\frac{\partial W_2}{\partial R} && R=R_2\,. \end{aligned} \tag{2.8}$$

From (2.7) and (2.8), we obtain

$$\begin{aligned} &-H_0R_1+\frac{I_0}{R_1^2}=-H_1R_1+\frac{I_1}{R_1^2} \\ &-H_0-\frac{2I_0}{R_1^3}=\mu\left(-H_1-\frac{2I_1}{R_1^3}\right) \end{aligned} \tag{2.9}$$

and

$$\begin{aligned} &-H_1R_2+\frac{I_1}{R_2^2}=-H_2R_2 \\ &\mu\left(-H_1-\frac{2I_1}{R_2^3}\right)=-H_2\,. \end{aligned} \tag{2.10}$$

Solving (2.9) and (2.10), we obtain

$$\begin{pmatrix} -H_0 \\ I_0 \end{pmatrix}=A_1\begin{pmatrix} -H_1 \\ I_1 \end{pmatrix} \tag{2.11}$$

and

$$\begin{pmatrix} -H_1 \\ I_1 \end{pmatrix} = A_1 \begin{pmatrix} -H_2 \\ 0 \end{pmatrix} \tag{2.12}$$

where A_1 and A_2 denote the following two matrices:

$$A_1 = \begin{pmatrix} \dfrac{\mu+2}{3} & \dfrac{2(1-\mu)}{3}\dfrac{1}{R_1^3} \\ \dfrac{1-\mu}{3}R_1^3 & \dfrac{2\mu+1}{3} \end{pmatrix} \tag{2.13}$$

$$A_2 = \begin{pmatrix} \dfrac{\mu^{-1}+2}{3} & \dfrac{2(1-\mu)^{-1}}{3}\dfrac{1}{R_2^3} \\ \dfrac{1-\mu^{-1}}{3}R_2^3 & \dfrac{2\mu^{-1}+1}{3} \end{pmatrix}. \tag{2.14}$$

From (2.11) and (2.12), we have

$$\begin{pmatrix} -H_0 \\ I_0 \end{pmatrix} = A_1 A_2 \begin{pmatrix} -H_2 \\ 0 \end{pmatrix}. \tag{2.15}$$

On calculating $A_1 A_2$,

$$S_i = H_2/H_0 \tag{2.16}$$

where

$$S_i = \frac{1}{1+\dfrac{2}{9}\dfrac{(\mu-1)^2}{\mu}\left(1-\dfrac{R_2^3}{R_1^3}\right)} \;{}^{*)} \tag{2.17}$$

which is hereafter called the shielding ratio.

Changes in S_i with μ are shown in Table 2.1 and in two ways in Figs. 2.2 and 2.3 for a few values of R_2/R_1. The figures show that for $\mu=1000$, R_2/R_1 should be 0.8 or smaller in order for the field in the inner space to be less than 1 % of the external-applied field.

*) S_i for a magnetic field expressed by a spherical harmonic of degree n becomes

$$S_i = \frac{1}{1-\dfrac{n(n+1)}{(2n+1)^2}\dfrac{(\mu-1)^2}{\mu}\left(1-\dfrac{R_2^{2n+1}}{R_1^{2n+1}}\right)}$$

TABLE 2.1 Magnetic shielding ratio S_i for a spherical shell.

R_2/R_1 / μ	0.9	0.8	0.4	0.0
1	1.00000	1.00000	1.00000	1.00000
5	0.83843	0.74238	0.60038	0.58442
10	0.67214	0.53237	0.37247	0.35714
50	0.25695	0.16110	0.09101	0.08568
100	0.14488	0.08600	0.04676	0.04390
500	0.03227	0.01818	0.00956	0.00896
1000	0.01637	0.00916	0.00477	0.00449
5000	0.00331	0.00184	0.00096	0.00090

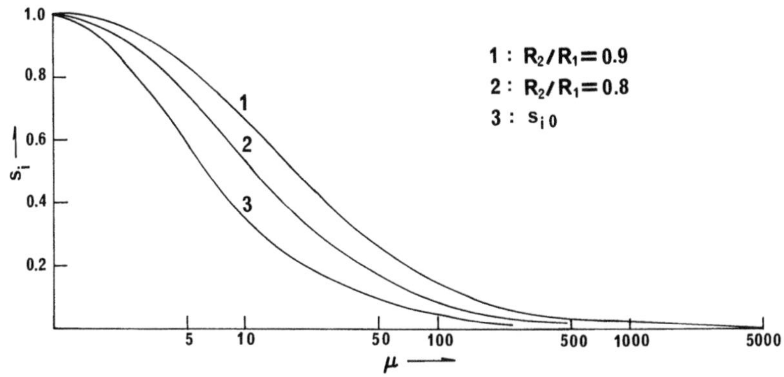

FIG. 2.2 Changes in the shielding ratio S_i for a spherical shell as the magnetic permeability increases. The curves 1 and 2 are for shells for which the ratio of outer radius R_1 to inner radius R_2 is 0.9 and 0.8, respectively. S_{i0} as shown by curve 3 gives the changes for the limiting case $R_2/R_1 \to 0$.

It is clear that S_i cannot be smaller than S_{i0} which is defined by S_i in the limit $R_2/R_1 \to 0$ and is

$$S_{i0} = \frac{1}{1+\frac{2}{9}\frac{(\mu-1)^2}{\mu}} = \frac{9\mu}{(2\mu+1)(\mu+2)}. \tag{2.18}$$

This gives the limit of the shielding effect for a given μ. Changes in S_{i0} are also shown in Fig. 2.2.

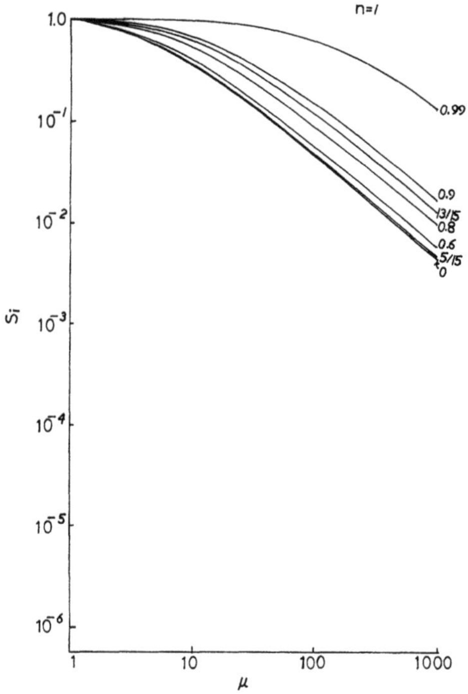

FIG. 2.3 Changes in shielding ratio S_i for a spherical shell with μ as expressed by its logarithm. R_2/R_1 is taken as a parameter, the values are attached to each curve in the figure.

2.2.1.2 Magnetic lines of force

Distribution in the xz plane for the three-dimensional case

Since the present problem is essentially a three-dimensional one, it is not easy to see how the magnetic lines of force behave around the shell. Nevertheless, it is of interest to examine the distribution of magnetic lines of force in a plane through the external magnetic field direction (the $\theta=0$ axis). Specifically we examine the behaviour of magnetic lines of force in the plane defined by $y=0$ or $\phi=0$. Cartesian coordinates (x, y, z) are related to spherical polar coordinates (R, θ, ϕ) in the following manner:

$$
\begin{aligned}
x &= R \sin\theta \cos\phi \\
y &= R \sin\theta \cos\phi \\
z &= R \cos\theta\,.
\end{aligned}
\tag{2.19}
$$

The R and θ components of the magnetic field in region 0 ($R>R_1$) are readily obtained from the first equation of (2.7) as

$$H_R=-\frac{\partial W_0}{\partial R}=\left(H_0+\frac{2I_0}{R^3}\right)\cos\theta$$
$$H_\theta=-\frac{1}{R}\frac{\partial W_0}{\partial\theta}=\left(-H_0+\frac{I_0}{R^3}\right)\sin\theta\,. \tag{2.20}$$

According to the definition of magnetic lines of force, we have

$$\frac{\mathrm{d}R}{H_R}=\frac{R\mathrm{d}\theta}{H_\theta} \tag{2.21}$$

so that

$$\frac{\mathrm{d}R}{\left(H_0+\frac{2I_0}{R^3}\right)\cos\theta}=\frac{R\mathrm{d}\theta}{\left(-H_0+\frac{I_0}{R^3}\right)\sin\theta} \tag{2.22}$$

which reduces to

$$\frac{I_0-H_0R^3}{2I_0+H_0R^3}\frac{\mathrm{d}R}{R}=\frac{\cos\theta}{\sin\theta}\,\mathrm{d}\theta\,. \tag{2.23}$$

On integrating (2.23), we obtain

$$\frac{1}{2}\log_e\frac{H_0R^3+2I_0}{R}+\log_e\sin\theta=\text{const.}$$

which can be rewritten

$$(R^3+2)\sin^2\theta=c^2R \tag{2.24}$$

if the parameters involved take the values:

$$\mu=\infty,\ H_0=1 \text{ and } R_1=1.$$

Equation (2.24), where c is a constant, is the equation of magnetic lines of force.

When R tends to infinity, (2.24) becomes

$$R\sin\theta=c \text{ or } x=c$$

so that the magnetic lines of force are parallel to the z axis.

Corresponding to c=0.5, 1.0, 1.5, 2.0, 2.5 and 3.0, Equation (2.24) specifies the curves which are shown in Fig. 2.4. Although only a solid sphere with an infinite permeability is considered here, the example gives an indication of the form of magnetic lines of force around a highly permeable sphere. It is of course possible to illustrate the distribution of magnetic lines of force for a permeable spherical shell after a somewhat more complicated calculation.

Two-dimensional approach

In contrast to the above case, let us study the magnetic lines of force around a circular cylinder of infinite length. The configuration of magnetic lines of force in a section perpendicular to the cylinder axis may have some similarity to the above case.

The problem is treated as two-dimensional in the xz plane. In that case, we can define a magnetic flux function ψ which is related to the magnetic flux density by

$$B_x=-\frac{\partial\psi}{\partial z},\quad B_z=\frac{\partial\psi}{\partial x} \tag{2.25}$$

and ψ satisfies

$$\frac{\partial^2\psi}{\partial x^2}+\frac{\partial^2\psi}{\partial z^2}=0. \tag{2.26}$$

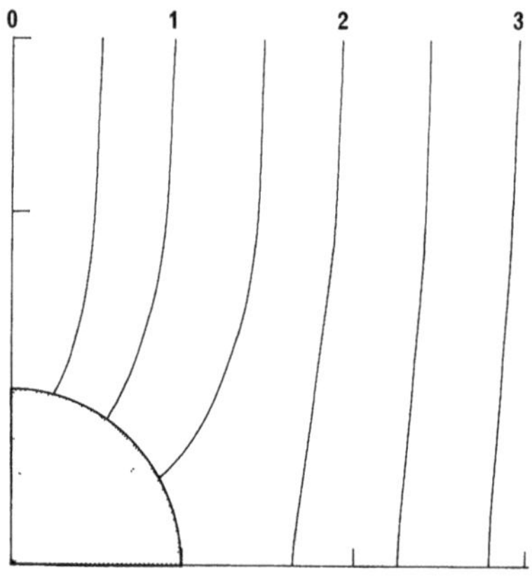

FIG. 2.4 Magnetic lines of force around a sphere having an infinitely large magnetic permeability. The figure shows the distribution of magnetic lines of force in a plane which is parallel to the external magnetic field and includes the center of the sphere.

An appropriate solution of (2.26) satisfying the boundary conditions can readily be provided. For the sake of simplicity, however, we shall study the case for which $\mu=\infty$. Then ψ for $r>r_1$ is given by

$$\psi=H_0\left(r+\frac{r_1^2}{r}\right)\sin\theta \tag{2.27}$$

where r_1 denotes the radius of the cylinder.

Polar coordinates (r, θ) are related to Cartesian coordinates by

$$x=r\sin\theta,\quad z=r\cos\theta. \tag{2.28}$$

The distribution of magnetic lines of force for this case with $H_0=1$ and $r_1=1$ can be seen in Fig. 2.5.

Axisymmetric approach

In contrast to the former two treatments, it is physically more appropriate to solve the problem taking into acount axial symmetry about the $\theta=0$ axis.

For axial symmetry about the z axis, we consider cylindrical coordinates (r, θ, z). Attention should be drawn to the point that this r is different from R for the spherical polar coordinates in the preceding statements. The equations to be satisfied by the magnetic potential and magnetic flux function are

$$\frac{\partial}{\partial r}\left(r\frac{\partial W}{\partial r}\right)+\frac{\partial}{\partial z}\left(r\frac{\partial W}{\partial z}\right)=0 \tag{2.29}$$

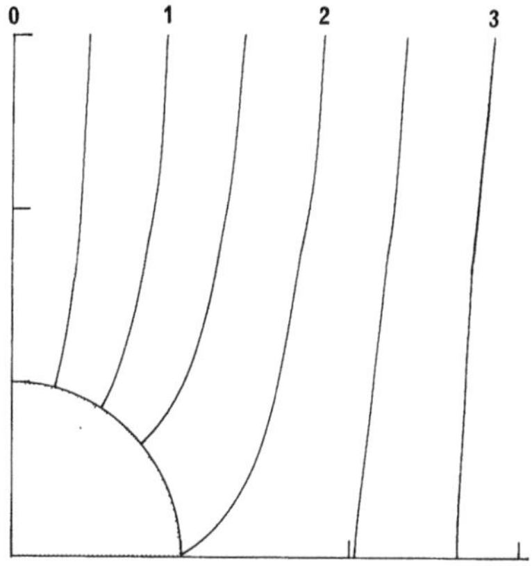

FIG. 2.5 Magnetic lines of force around an infinitely long cylinder having an infinitely large magnetic permeability in a cross section perpendicular to the cylinder axis. The cross section plane is parallel to the external magnetic field.

and

$$\frac{\partial}{\partial r}\left(\frac{1}{r}\frac{\partial\psi}{\partial r}\right)+\frac{\partial}{\partial z}\left(\frac{1}{r}\frac{\partial\psi}{\partial z}\right)=0. \tag{2.30}$$

Equations similar to (2.25) together with

$$H_r=-\frac{\partial W}{\partial r}=-\frac{1}{r}\frac{\partial\psi}{\partial z},\quad H_z=-\frac{\partial W}{\partial z}=\frac{1}{r}\frac{\partial\psi}{\partial r} \tag{2.31}$$

imply

$$\frac{\partial W}{\partial r}\frac{\partial\psi}{\partial r}+\frac{\partial W}{\partial z}\frac{\partial\psi}{\partial z}=0, \tag{2.32}$$

so that the curves for W=const. and ψ=const. are orthogonal.

If we adopt a magnetic flux function for the axisymmetric problem relating to a shell as discussed above, the magnetic flux functions for the respective regions become

$$\psi_0=\left(\frac{H_0}{2}r^2+\frac{I_0}{r}\right)\sin^2\theta \qquad r>r_1$$

$$\psi_1=\mu\left(\frac{H_1}{2}r^2+\frac{I_1}{r}\right)\sin^2\theta \qquad r_1>r>r_2 \tag{2.33}$$

and

$$\psi_2=\frac{H_2}{2}r^2\sin^2\theta \qquad r_2>r$$

where r_1 and r_2 denote the outer and inner radii of the shell, respectively.

When $\mu=\infty$, we obtain

$$\psi_0=H_0\left(\frac{1}{2}r^2+\frac{r_1^{\,3}}{r}\sin^2\theta\right). \tag{2.34}$$

Assuming $H_0=1$ and $r_1=1$, the magnetic lines of force based on (2.34) are illustrated in Fig. 2.6 in which it should be noted that the interval between neighbouring magnetic lines of force becomes smaller as r increases.

In the following discussion of multiple shells, the magnetic flux function in the framework of axisymmetry will be used although the actual distribution of magnetic lines of force in a cross section such as the xz plane is somewhat different from that shown in the rz plane for the axisymmetric case. When we compare Fig. 2.6 to Fig. 2.4, it is seen that even the distribution of magnetic lines of force for the axisymmetric

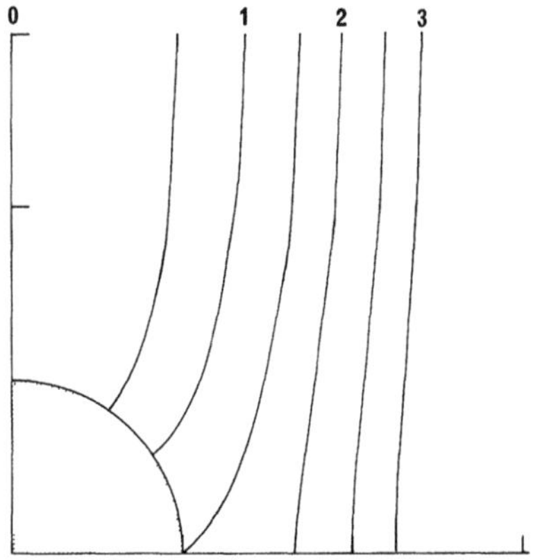

FIG. 2.6 Magnetic lines of force around a sphere having an infinitely large magnetic permeability, expressed by lines along with the magnetic flux function defined by the axisymmetric coordinate system takes on a constant value.

case can provide an approximate distribution for the actual one in a cross section.

2.2.1.3 Thick and thin shells

In order to see the difference in the shielding effect between thick and thin shells, the distribution of magnetic lines of force and magnetic flux densities are calculated for $R_2/R_1=5/15$ and $13/15$, while μ is assumed as 100 and 1000. The results are shown in Figs. 2.8 to 2.15. Figure 2.7 shows the magnetic lines of force for an external magnetic field of uniform distribution parallel to the z axis. As stated before, the interval between the neighbouring fluxes becomes small as r or y increases.

Although it is seen that the thicker the shell becomes, the smaller becomes the shielding ratio, the deflection of magnetic flux is greater for a thin shell. It is therefore unwise to make the thickness large in order to achieve a high shielding effect. Judging from the fact that the bending of magnetic flux is most pronounced along the inner boundary of the shell, we may see that more efficient shielding can be achieved by making use of multiple shells as discussed in a later section.

2.2.1.4 A magnetic dipole in a shell—an external problem

It is a matter of theoretical as well as practical interest to see to what extent the magnetic field of a magnet will be shielded by a magnetic shell surrounding it. We shall here study a magnetic dipole placed at the center of a spherical shell of which the outer and inner radii are denoted by R_1 and R_2, respectively.

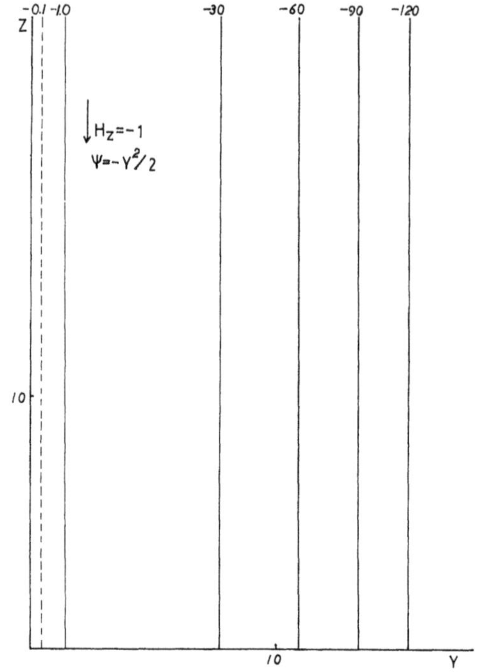

FIG. 2.7 Magnetic lines of force of a uniform external magnetic field in the z direction. Because of the axisymmetric coordinates, the interval between the neighbouring lines becomes small as r or y increases.

Assuming that the dipole, of magnetic moment M, is directed along the z axis, the magnetic potentials in regions 0 ($R>R_1$), 1 ($R_1>R>R_2$) and 2 ($R_2>R$) may be written as

$$\begin{aligned} W_0 &= \frac{I_0}{R^2}\cos\theta && R>R_1 \\ W_1 &= \left(H_1R+\frac{I_1}{R^2}\right)\cos\theta && R_1>R>R_2 \\ W_2 &= \left(\frac{M}{R^2}+H_2R\right)\cos\theta && R_2>R\,. \end{aligned} \tag{2.35}$$

The boundary conditions which are the same as those given by Eq. (2.8) imply

$$\begin{aligned} \frac{I_0}{R_1^2} &= R_1H_1+\frac{I_1}{R_1^{\,2}} \\ \frac{-2I_0}{R_1^{\,3}} &= \mu\left(H_1-\frac{2I_1}{R_1^{\,3}}\right) \end{aligned} \tag{2.36}$$

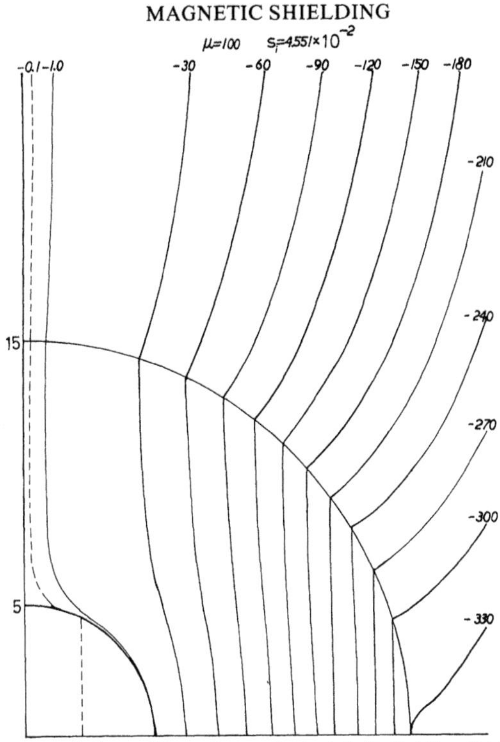

FIG. 2.8 Magnetic lines of force for a spherical shell. $R_2/R_1=5/15$ and $\mu=100$ are assumed.

$$
\begin{aligned}
R_2H_1+\frac{I_1}{R_2{}^2}&=R_2H_2+\frac{M}{R_2{}^2}\\
\mu\left(H_1-\frac{2I_1}{R_2{}^3}\right)&=H_2-\frac{2M}{R_2{}^3}\,.
\end{aligned}
\tag{2.37}
$$

I_0, H_1, I_1 and H_2 can be determined from Eqs. (2.36) and (2.37) as

$$
\begin{aligned}
I_0&=\frac{9\mu M}{(\mu+2)(2\mu+1)-2(\mu-1)^2\left(\dfrac{R_2}{R_1}\right)^3}\\
H_1&=\frac{2(\mu-1)}{3\mu}\,\frac{I_0}{R_1{}^3}\\
I_1&=\frac{\mu+2}{3\mu}\,I_0\\
H_2&=-\frac{2(\mu-1)(\mu+2)\dfrac{1}{R_2{}^3}\left(1-\dfrac{R_2{}^3}{R_1{}^3}\right)}{(\mu+2)(2\mu+1)-2(\mu-1)^2\left(\dfrac{R_2}{R_1}\right)^3}\,M\,.
\end{aligned}
\tag{2.38}
$$

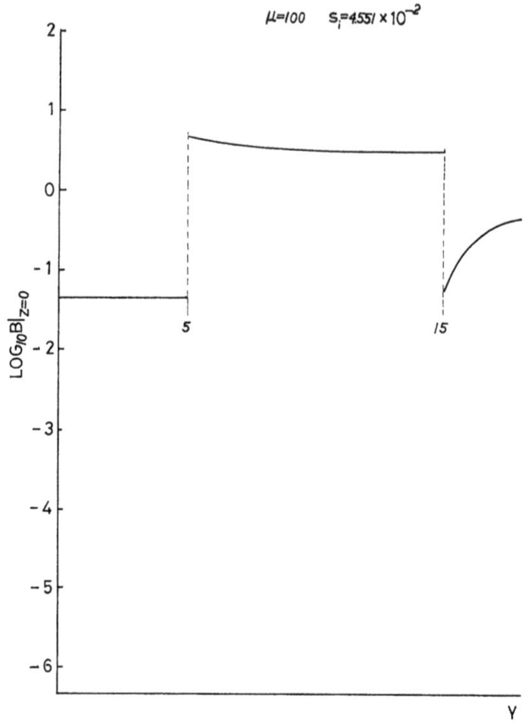

FIG. 2.9 Logarithmic z component of magnetic flux density in the $z=0$ plane. $R_2/R_1=5/15$ and $\mu=100$ are assumed.

It is therefore seen that a dipole field appears outside the shell. The field is weaker than it would be in the absence of the shell, by a factor S_e which is given by

$$S_e=\frac{9\mu}{(\mu+2)(2\mu+1)-2(\mu-1)^2\left(\frac{R_2}{R_1}\right)^3} \tag{2.39}$$

which may also be called the shielding ratio as before.

Equation (2.39) can be rewritten as

$$S_e=\frac{1}{1+\frac{2}{9}\frac{(\mu-1)^2}{\mu}\left(1-\frac{R_2^3}{R_1^3}\right)} \tag{2.40}$$

so that we see that $S_i=S_e(=S)$ holds good as far as n is the same. What we have learned in Subsection 2.2.1. 1 about the degree of shielding to

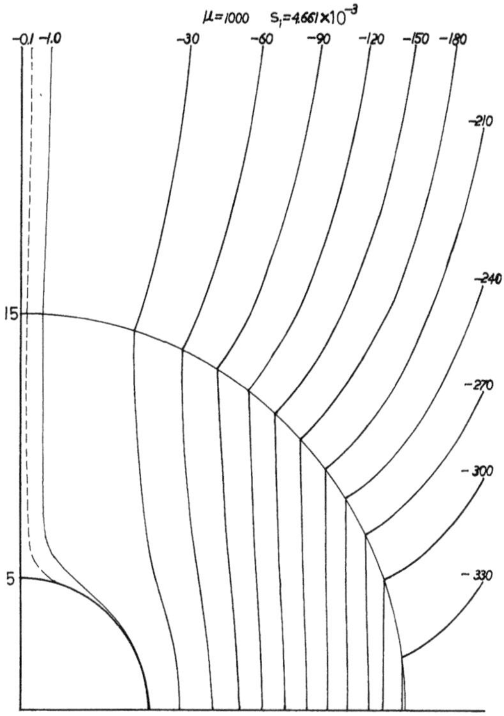

FIG. 2.10 Magnetic lines of force for a spherical shell. $R_2/R_1=5/15$ and $\mu=1000$ are assumed.

an external field is then applicable to the shielding of the internal dipole. Figure 2.16 shows how S changes with an increase in R_2/R_1 taking μ as a parameter, the figure of course also holding good for the shielding of the external magnetic field as studied in Subsection 2.2.1.

It is also obvious that the relation $S_i=S_e$ holds good for any magnetic field expressed by a spherical harmonic of degree n.

2.2.1.5 Magnetic field arising from a shell having a remanent magnetization

It is of theoretical (and perhaps practical) interest to determine the magnetic field produced by a magnetized spherical shell although such a problem has nothing to do with the problem of shielding. The results of the present study may be applied to the magnetic field arising from the moon's crust which was magnetized by an external or internal magnetic field some time in the past. The magnetizing field is presumed to have now disappeared.

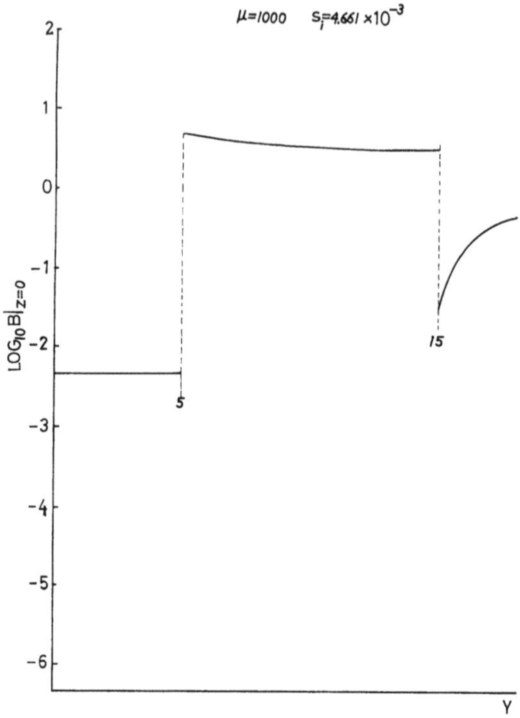

FIG. 2.11 Logarithmic z component of magnetic flux density in the z=0 plane. $R_2/R_1=5/15$ and $\mu=1000$ are assumed.

A shell magnetized by an external field

The magnetization in the shell as denoted by $\boldsymbol{J}$ is given by

$$\boldsymbol{J}=\kappa \boldsymbol{H} \tag{2.41}$$

where $\boldsymbol{H}$ is the magnetizing field and κ the magnetic susceptibility.

When the magnetizing field is a uniform one external to the shell as shown in Fig. 2.1, $\boldsymbol{J}$ is given by

$$\boldsymbol{J}=-\kappa H_0 \operatorname{grad}(R\cos\theta) \tag{2.42}$$

where H_0 is the intensity of the magnetizing field parallel to the $\theta=0$ axis. Spherical polar coordinates (R, θ, ϕ) are used here.

An analogy between the magnetized shell and a bar magnet leads to the point that the magnetic field due to the shell can be calculated by

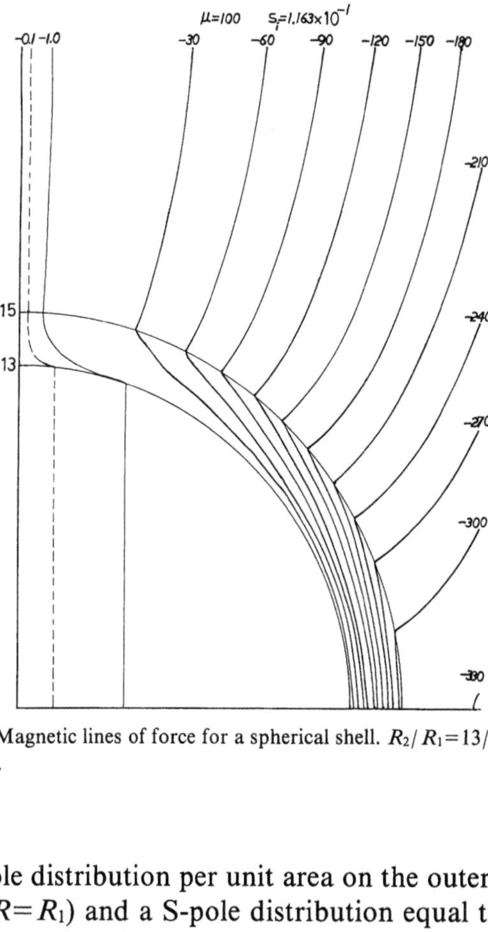

FIG. 2.12 Magnetic lines of force for a spherical shell. $R_2/R_1=13/15$ and $\mu=100$ are assumed.

taking a N-pole distribution per unit area on the outer surface $R=R_1$ equal to J_R $(R=R_1)$ and a S-pole distribution equal to $-J_R$ $(R=R_2)$ over the inner surface $R=R_2$. The volume distribution of magnetic charge within the shell is everywhere zero because div $\boldsymbol{J}=0$ as can readily be deduced from (2.42). Hence the problem reduces to obtaining the magnetic field due to the above-mentioned surface distributions of magnetic poles.

Returning to Fig. 2.1, the magnetic potentials in regions 0, 1 and 2 are respectively given as

$$\begin{aligned} W_0&=A_1\frac{\cos\theta}{R^2} && R>R_1 \\ W_1&=\left(B_1R+\frac{C_1}{R^2}\right)\cos\theta && R_1>R>R_2 \\ W_2&=D_1R\cos\theta && R_2>R \end{aligned} \qquad (2.43)$$

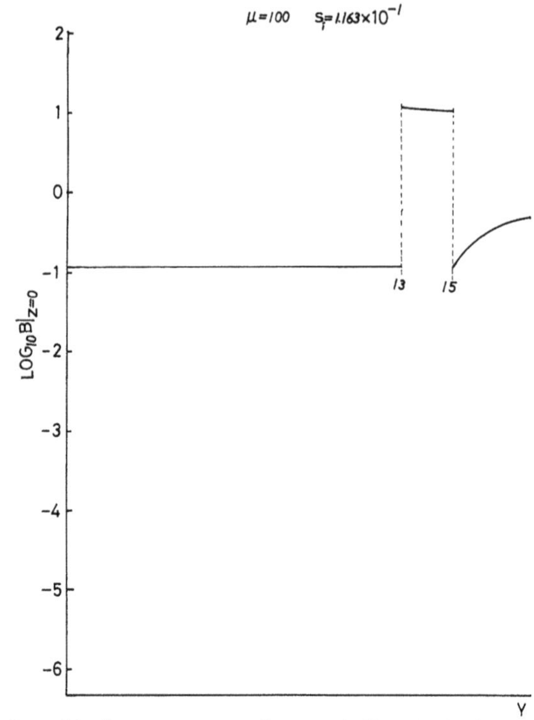

FIG. 2.13 Logarithmic z component of magnetic flux density in the $z=0$ plane. $R_2/R_1=13/15$ and $\mu=100$ are assumed.

where A_1, B_1, C_1 and D_1 are constants to be determined by the boundary conditions.

According to the theory of magnetism, a jump amounting to $4\pi\sigma$ should occur in the normal magnetic field component across a spherical surface over which the magnetic charge distribution is given by σ. It is therefore seen that the boundary conditions applicable at $R=R_1$ and $R=R_2$ are the continuity of potential and the discontinuity of the normal magnetic field component by the above amount.

The following equations are then obtained:

$$\begin{aligned}
&\frac{A_1}{R_1^3}-B_1-\frac{C_1}{R_1^3}=0\\
&\frac{2A_1}{R_1^3}+B_1-\frac{2C_1}{R_1^3}=-4\pi\kappa H_0\\
&B_1+\frac{C_1}{R_2^3}-D_1=0\\
&-B_1+\frac{2C_1}{R_2^3}+D_1=4\pi\kappa H_0
\end{aligned} \tag{2.44}$$

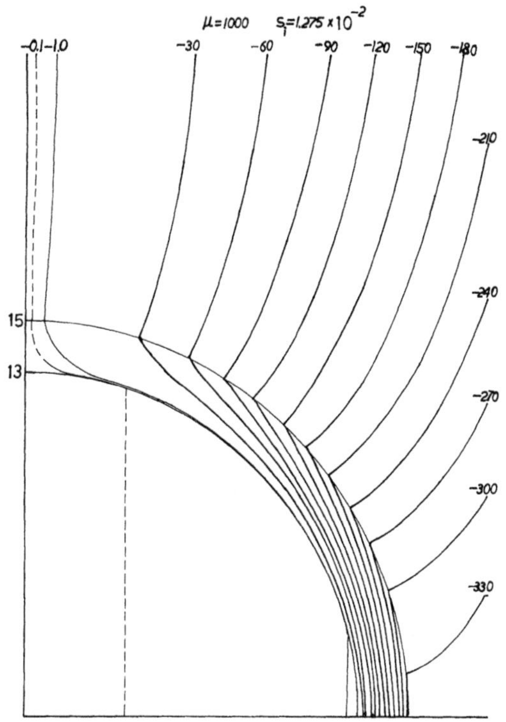

FIG. 2.14 Magnetic lines of force for a spherical shell. $R_2/R_1=13/15$ and $\mu=1000$ are assumed.

solving which we obtain

$$\begin{aligned} A_1 &= -\frac{4\pi}{3}\kappa\,(R_1^3-R_2^3)\,H_0 \\ B_1 &= -\frac{4\pi}{3}\kappa H_0 \\ C_1 &= \frac{4\pi}{3}\kappa R_2^3 H_0 \\ D_1 &= 0\,. \end{aligned} \tag{2.45}$$

It is thus proved that the shell has no field internal to itself. RUNCORN (1975) extended the above theory to a spherical shell magnetized by an external magnetic field of which the magnetic potential is given by

$$V=R_1 q_n \left(\frac{R}{R_1}\right)^n S_n\,(\theta,\,\phi) \tag{2.46}$$

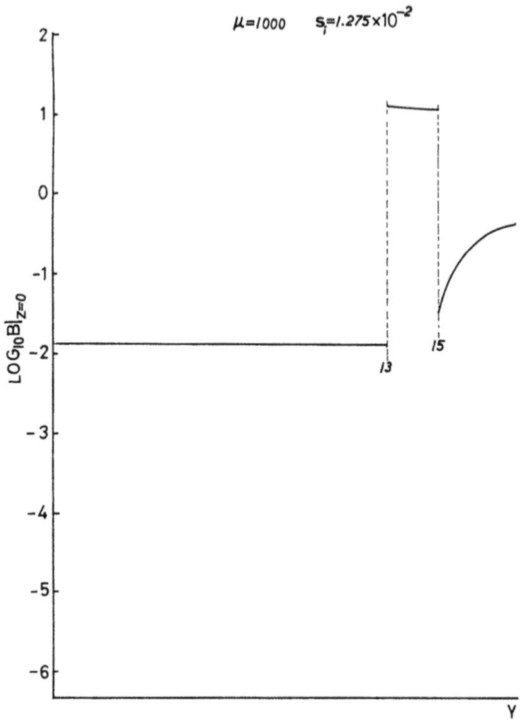

FIG. 2.15 Logarithmic z component of magnetic flux density in the $z=0$ plane. $R_2/R_1=13/15$ and $\mu=1000$ are assumed.

where S_n is a surface harmonic of degree n and q_n a constant proportional to the intensity of the external inducing field.

The magnetic potentials in the respective regions are given as

$$
\begin{aligned}
W_0 &= A_n R_1 \left(\frac{R_1}{R}\right)^{n+1} S_n & R &> R_1 \\
W_1 &= R_1 \left\{ B_n \left(\frac{R}{R_1}\right)^n + C_n \left(\frac{R_1}{R}\right)^{n+1} \right\} S_n & R_1 &> R > R_2 \qquad (2.47)\\
W_2 &= D_n R_1 \left(\frac{R}{R_1}\right)^n S_n & R_2 &> R\,.
\end{aligned}
$$

From the boundary conditions, the constants can be determined as

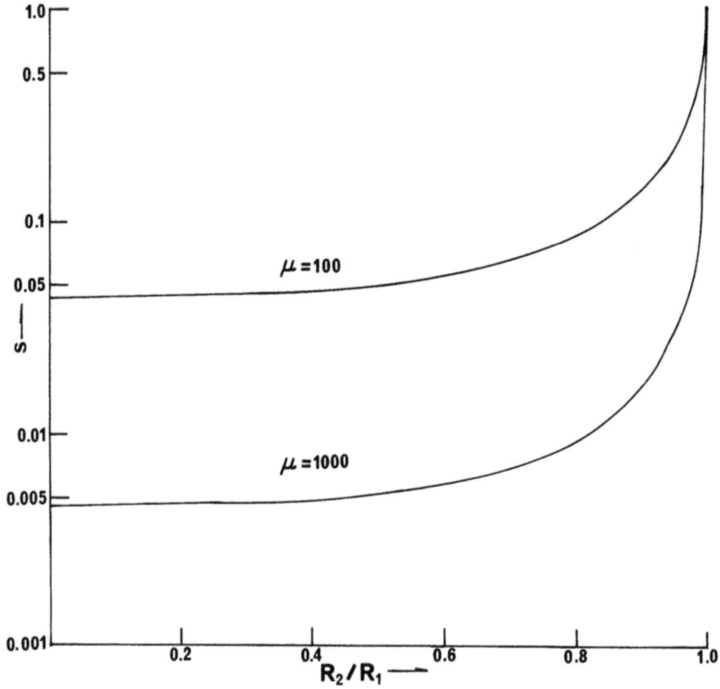

FIG. 2.16 Changes in the shielding ratio S ($S_i=S_e$) with the thickness of the spherical shell. Two values, 100 and 1000 are assigned to magnetic permeability μ.

$$\begin{aligned}
A_n &= -4\pi\kappa q_n \frac{n}{2n+1}\left\{1-\left(\frac{R_2}{R_1}\right)^{2n+1}\right\} \\
B_n &= -4\pi\kappa q_n \frac{n}{2n+1} \\
C_n &= 4\pi\kappa q_n \frac{n}{2n+1}\left(\frac{R_2}{R_1}\right)^{2n+1} \\
D_n &= 0.
\end{aligned} \tag{2.48}$$

Putting $n=1$, we see that the equations in (2.48) reduce to those in (2.45). It is interesting to note that a spherical shell magnetized by any magnetic field external to it produces no magnetic field internal to itself.

Magnetic lines of force for $\boldsymbol{H}$ can be obtained based on an equation similar to (2.21). Lines of force for $\boldsymbol{B}$ are also obtained by taking the relation

$$B = H + 4\pi J \tag{2.49}$$

into account.

Figure 2.17 (RUNCORN, 1975) shows examples of lines of ***B*** and ***H*** for cases in which $n=1$.

A shell magnetized by an internal field

The above theory can readily be modified in such a manner that it is applicable to obtaining the magnetic field arising from a shell magnetized by an internal magnetic field.

Following RUNCORN (1975), let us assume that the inducing magnetic field is given by

$$V = R_1 g_n \left(\frac{R_1}{R}\right)^{n+1} S_n(\theta, \phi) \tag{2.50}$$

where g_n is proportional to the intensity of the inducing field.

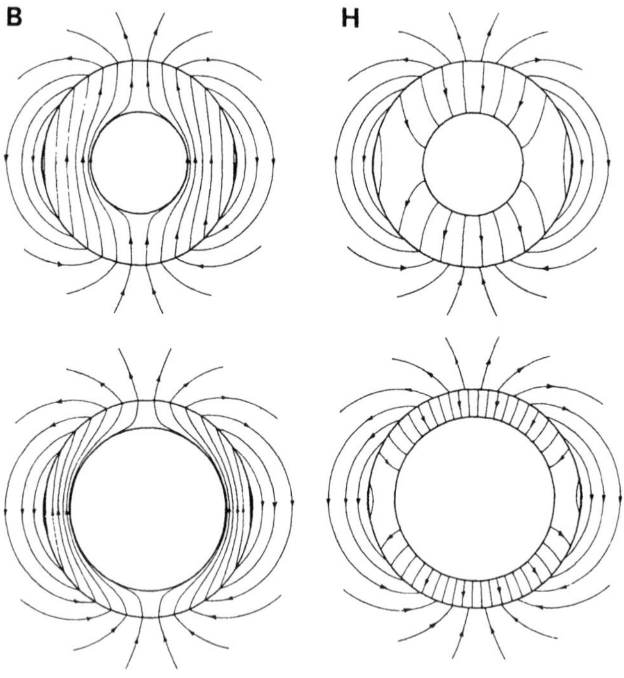

FIG. 2.17 Lines of ***B*** and ***H*** for a spherical shell magnetized by an external uniform magnetic field, which has now vanished. (The magnetic field magnetizing the thinner shell is 1.5 times the other) (RUNCORN, 1975).

The boundary conditions at $R=R_1$ and $R=R_2$ lead to equations similar to Eq. (2.44) from which we obtain

$$\begin{aligned} &A_n=0 \\ &B_n=-4\pi\kappa g_n\frac{n+1}{2n+1} \\ &C_n=4\pi\kappa g_n\frac{n+1}{2n+1} \\ &D_n=-4\pi\kappa g_n\frac{n+1}{2n+1}\left\{1-\left(\frac{R_1}{R_2}\right)^{2n+1}\right\}. \end{aligned} \tag{2.51}$$

With these constants, we can determine the magnetic potentials as given in Eq. (2.47) for the respective regions.

It is therefore evident that no field appears outside the shell. RUNCORN (1975) presented some interesting drawings as reproduced in Fig. 2.18 in which the lines of ***B*** and ***H*** are shown for thick and thin shells magnetized by an internal dipole field.

This fact has an important bearing on lunar magnetism. Supposing that dynamo action had been in operation in the moon's

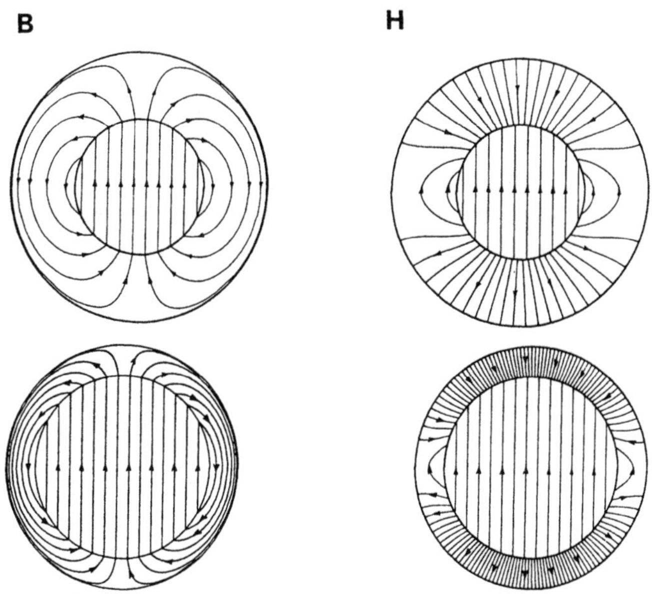

FIG. 2.18 Lines of ***B*** and ***H*** for a spherical shell magnetized by an internal (dipole) magnetic field, which has now vanished. (The dipole magnetizing the thinner shell is 5 times the other) (RUNCORN, 1975).

interior some time in the past, it is not unreasonable to suppose that the outer layer of the moon became magnetized by the magnetic field arising from the dynamo. If the dynamo action ceased at a certain time, the magnetic field on the lunar surface would disappear because $A_n=0$ although the outer layer would remain magnetized.

Should a portion of the lunar crust be broken by a meteor collision or some other cause, a local magnetic field would leak outside the superficial layer of the moon because of the magnetization in the layer.

It has been made clear that the overall magnetic field of the moon is very small. According to the results obtained by the Apollo 16 subsatellite, the lunar magnetic moment is smaller than the earth's magnetic moment by a factor of 10^{-8} or thereabout. However, a magnetic field reaching 2.5 nT was observed at an altitude of 100 km above the lunar surface. The astronauts on board the Apollo 12 lunar module conducted a magnetic survey on the lunar surface by means of a portable magnetometer and found a local magnetic field amounting to as much as 320 nT.

On the other hand, magnetic properties of rock samples brought back to the earth by the Apollo missions have been extensively studied. All through the studies, it has become apparent that many lunar rock samples have stable natural remanent magnetization of the order of 10^{-5} e.m.u./gm.

Summarizing all these observed facts, RUNCORN (1975) proposed that the moon's outer layer was magnetized some time in the past when dynamo action was taking place within the moon. The dynamo action having stopped later, no general magnetic field outside the moon is now observed although the moon's outer layer is still magnetized. If a portion of the moon's crust is broken by a meteor collision or some other cause, a magnetic field leaks out over the nearby lunar surface producing a local anomalous magnetic field. It appears that the most outstanding magnetic anomalies on the lunar surface are correlated with large craters.

2.2.2 *Multiple shells*

In Subsection 2.2.1.3, it was seen that the bending of magnetic lines of force is most pronounced along the inner boundary of a shell as demonstrated in some figures in that subsection. Accordingly, it is suggested that the shielding to an external field may be more effective with a system consisting of a number of shells than for a single thick shell. It is the purpose of this subsection, therefore, to study the shielding effect of multiple spherical shells.

2.2.2.1 Theory for multiple shells

Let us take a model consisting of n shells as shown in Fig. 2.19, the system being placed in a uniform external field H_0 parallel to the $\theta=0$ axis as is the case of a single shell studied in Subsection 2.2.1.1.

Magnetic potentials, constants to be determined by the boundary conditions and the radii of shell boundaries are defined in a manner similar to the case of a single shell. The magnetic potentials in the respective regions are then given as

$$\begin{aligned}
W_0 &= \left(-H_0 R + \frac{I_0}{R^2}\right)\cos\theta & R &> R_1 \\
W_1 &= \left(-H_1 R + \frac{I_1}{R^2}\right)\cos\theta & R_1 &> R > R_2 \\
&\vdots \\
W_j &= \left(-H_j R + \frac{I_j}{R^2}\right)\cos\theta & R_j &> R > R_{j+1} \\
&\vdots \\
W_{2n} &= -H_{2n} R\cos\theta & R_{2n} &> R
\end{aligned} \tag{2.52}$$

in which

$$j=0,\ 1,\ 2,\ \cdots,\ 2n\,.$$

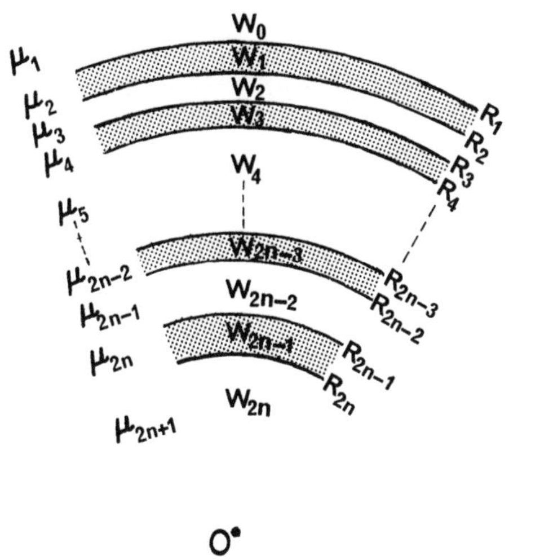

FIG. 2.19 Spherical shell model consisting of n shells.

It is seen that the magnetic field inside the innermost shell is uniform and parallel to the external field.

On introducing μ_j^* which is defined by

$$\mu_j^* = \mu_{j+1}/\mu_j \ (j=1, 2, \cdots, 2n) \tag{2.53}$$

the boundary conditions may be written as

$$\begin{aligned}
&W_0 = W_1, \quad \frac{\partial W_0}{\partial R} = \mu_1^* \frac{\partial W_1}{\partial R} && R=R_1 \\
&W_1 = W_2, \quad \frac{\partial W_1}{\partial R} = \mu_2^* \frac{\partial W_2}{\partial R} && R=R_2 \\
&\vdots \\
&W_{j-1} = W_j, \quad \frac{\partial W_{j-1}}{\partial R} = \mu_j^* \frac{\partial W_j}{\partial R} && R=R_j \\
&W_j = W_{j+1}, \quad \frac{\partial W_j}{\partial R} = \mu_{j+1}^* \frac{\partial W_{j+1}}{\partial R} && R=R_{j+1} \\
&\vdots \\
&W_{2n-1} = W_{2n}, \quad \frac{\partial W_{2n-1}}{\partial R} = \mu_{2n}^* \frac{\partial W_{2n}}{\partial R} && R=R_{2n}\,.
\end{aligned} \tag{2.54}$$

Putting (2.52) into (2.54), we obtain

$$\begin{aligned}
&-H_0R_1 + \frac{I_0}{R_1^2} = -H_1R_1 + \frac{I_1}{R_1^2} \\
&-H_0 - 2\frac{I_0}{R_1^3} = \mu_1^* \left(-H_1 - 2\frac{I_1}{R_1^3}\right) \\
&-H_1R_2 + \frac{I_1}{R_2^2} = -H_2R_2 + \frac{I_2}{R_2^2} \\
&-H_1 - 2\frac{I_1}{R_2^3} = \mu_2^* \left(-H_2 - 2\frac{I_2}{R_2^3}\right) \\
&\vdots \\
&-H_{j-1}R_j + \frac{I_{j-1}}{R_j^2} = -H_jR_j + \frac{I_j}{R_j^2} \\
&-H_{j-1} - 2\frac{I_{j-1}}{R_j^3} = \mu_j^* \left(-H_j - 2\frac{I_j}{R_j^3}\right) \\
&-H_jR_{j+1} + \frac{I_j}{R_{j+1}^2} = -H_{j+1}R_{j+1} + \frac{I_{j+1}}{R_{j+1}^2}
\end{aligned} \tag{2.55}$$

$$-H_j-2\frac{I_j}{R_{j+1}{}^3}=\mu_{j+1}{}^*\left(-H_{j+1}-2\frac{I_{j+1}}{R_{j+1}{}^3}\right)$$

$$\vdots$$

$$-H_{2n-1}R_{2n}+\frac{I_{2n-1}}{R_{2n}{}^2}=-H_{2n}R_{2n}$$

$$-H_{2n-1}-2\frac{I_{2n-1}}{R_{2n}{}^3}=-\mu_{2n}{}^*H_{2n}$$

where

$$j=1,\ 2,\ \cdots,\ 2n\,.$$

In a fashion similar to (2.11) and (2.12), (2.55) can be written as

$$\begin{pmatrix}-H_0\\ I_0\end{pmatrix}=A_1\begin{pmatrix}-H_1\\ I_1\end{pmatrix}$$

$$\begin{pmatrix}-H_1\\ I_1\end{pmatrix}=A_2\begin{pmatrix}-H_2\\ I_2\end{pmatrix}$$

$$\vdots$$

$$\begin{pmatrix}-H_{j-1}\\ I_{j-1}\end{pmatrix}=A_j\begin{pmatrix}-H_j\\ I_j\end{pmatrix} \tag{2.56}$$

$$\begin{pmatrix}-H_j\\ I_j\end{pmatrix}=A_{j+1}\begin{pmatrix}-H_{j+1}\\ I_{j+1}\end{pmatrix}$$

$$\vdots$$

$$\begin{pmatrix}-H_{2n-1}\\ I_{2n-1}\end{pmatrix}=A_{2n}\begin{pmatrix}-H_{2n}\\ 0\end{pmatrix}$$

where

$$A_j=\begin{pmatrix}\dfrac{\mu_j{}^*+2}{3} & \dfrac{2(1-\mu_j{}^*)}{3}\dfrac{1}{R_j^3}\\ \dfrac{1-\mu_j{}^*}{3}R_j^3 & \dfrac{2\mu_j{}^*+1}{3}\end{pmatrix}\ (j=1,\ 2,\ \cdots,\ 2n). \tag{2.57}$$

Therefore, we finally reach a relation

$$\begin{pmatrix}-H_0\\ I_0\end{pmatrix}=A_1A_2\cdots A_{2n}\begin{pmatrix}-H_{2n}\\ 0\end{pmatrix}. \tag{2.58}$$

In a fashion similar to (2.16), we may define the shielding ratio by

$$S_i = H_{2n}/H_0 = A_1^{-1}A_2^{-1}\cdots A_{2n}^{-1}. \tag{2.59}$$

When we assume that the magnetic permeabilities for all the shells are equal to μ and that the permeability in free space is unity, (2.57) becomes

$$A_{2i-1} = \begin{pmatrix} \frac{\mu+2}{3} & \frac{2(1-\mu)}{3}\frac{1}{R_{2i-1}^3} \\ \frac{1-\mu}{3}R_{2i-1}^3 & \frac{2\mu+1}{3} \end{pmatrix} \tag{2.60}$$

$$A_{2i} = \begin{pmatrix} \frac{\mu^{-1}+2}{3} & \frac{2(1-\mu)^{-1}}{3}\frac{1}{R_{2i}^3} \\ \frac{1-\mu^{-1}}{3}R_{2i}^3 & \frac{2\mu^{-1}+1}{3} \end{pmatrix} \tag{2.61}$$

where

$$i = 1, 2, \cdots, n.$$

These are the generalized expressions of (2.13) and (2.14).

It is thus possible to estimate the shielding ratio for multiple shells from (2.59) along with (2.60) and (2.61).

2.2.2.2 How efficient is a system of multiple shells?

In order to see that the magnetic shielding becomes more efficient as the number of shells increases, the shielding ratios are calculated for the four models shown in Fig. 2.20 in which the number of shells n takes

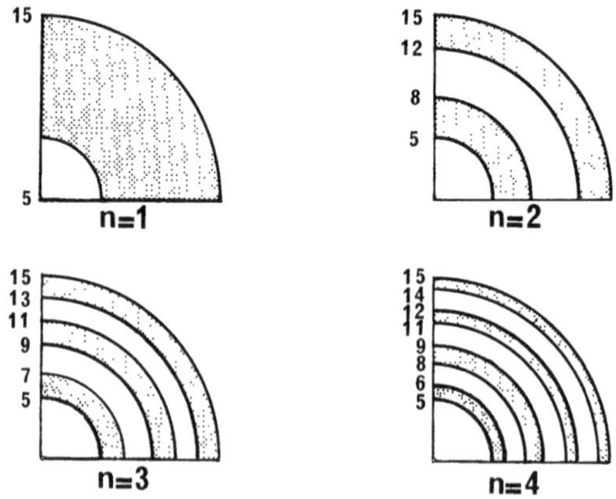

FIG. 2.20 Spherical shell models consisting of 1, 2, 3 and 4 shells. The outermost and innermost radii are common for all the models.

values from 1 to 4. The outermost and innermost radii are common for all the models.

Figure 2.21 shows how S_i changes with μ for the four models. It is clearly seen in the figure that the shielding becomes extremely high as n becomes large, so that a system of multiple shells is in general superior to a single shell as far as the shielding effect is concerned. The mass of magnetic matter for the models becomes smaller as n increases as can be seen in Fig. 2.20, so that multiple shells are advantageous from the standpoint of economy, too.

The tendency for the shielding to become more efficient as n increases seems to be even stronger for higher μ. In order to see this point more clearly, the shielding ratios are calculated for a single thick shell for which the radii are $R_1=300$ and $R_2=100$ and a model consisting of 5 extremely thin shells for which the radii are $R_1=300$, $R_2=299$, $R_3=251$, $R_4=250$, $R_5=201$, $R_6=200$, $R_7=151$, $R_8=150$, $R_9=101$ and $R_{10}=100$. The models are shown in Fig. 2.22.

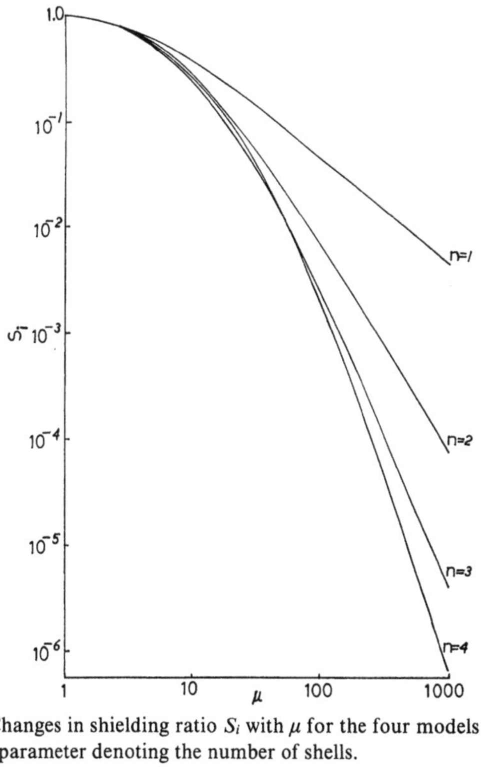

FIG. 2.21 Changes in shielding ratio S_i with μ for the four models shown in Fig. 2.20. n is the parameter denoting the number of shells.

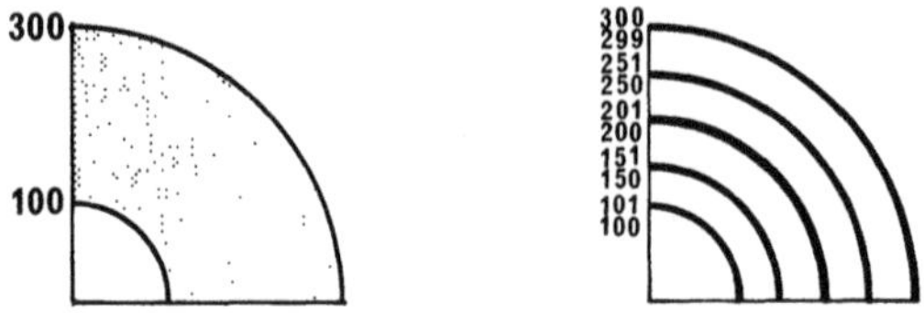

FIG. 2.22 A very thick single shell model and a 5-shell model of extremely small thickness.

Changes in S_i for both models are shown in Fig. 2.23 in which we see that the shielding is high for the single shell ($n=1$) when μ is small. However, the shielding effect of multiple shells ($n=5$) becomes larger than that for $n=1$ with μ's larger than 900 or so.

For $\mu=1000$, the $n=5$ model achieves a shielding effect 2.6 times that for the $n=1$ model. Meanwhile, the mass of magnetic matter of the former model amounts to one thirty-eighth of that for the latter model.

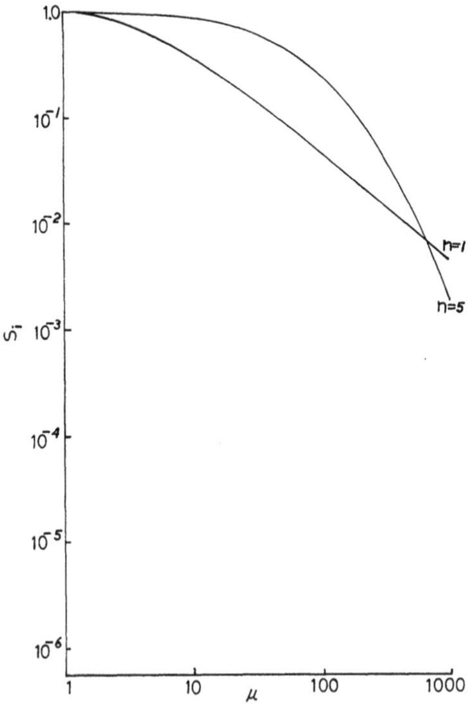

FIG. 2.23 Changes in the shielding ratio S_i with μ for the thick single shell model ($n=1$) and thin 5-shell model.

For many highly permeable substances, it is known that the effective permeability becomes higher as the magnetic field becomes weaker. It is therefore seen that use of a system of multiple shells is highly advantageous and economical for magnetic shielding.

2.2.2.3 Magnetic lines of force and their density

Solving the simultaneous equations given in (2.56), we can determine $I_0, H_1, I_1, \cdots, H_{2n-1}, I_{2n-1}$, and H_{2n} in terms of H_0 although H_{2n} has already been obtained in (2.58). Since the magnetic potential, magnetic flux function and magnetic flux density on the $z=0$ plane are given by

$$W_j=\left(-H_j r+\frac{I_j}{r^2}\right)\cos\theta \tag{2.62}$$

$$\psi_j=\mu_{j+1}\left(\frac{H_j}{2}r^2+\frac{I_j}{r}\right)\sin^2\theta \tag{2.63}$$

$$B_{j,\,z=0}=\mu_{j+1}\left(-H_j+\frac{I_j}{y^3}\right) \tag{2.64}$$

for the jth shell or free space as shown in Fig. 2.19, it is possible to calculate the above three quantities for each shell and for the free space. In this study, we use cylindrical coordinates (r, θ, z) taking axial symmetry into account. In the $z=0$ plane, we have relations

$$x=r\cos\theta,\quad y=r\sin\theta\,.$$

Figures 2.24, 2.25, 2.26 and 2.27 show the magnetic lines force and the magnetic flux density in the $z=0$ plane for $\mu=100$ and 1000. We assume $n=3$, $r_1=15$, $r_2=13$, $r_3=11$, $r_4=9$, $r_5=7$ and $r_6=5$ here. It is remarkable to see in these figures that the magnetic lines of force which enter the highly permeable shells are deflected along the inner boundaries, so that very few lines reach the innermost space. That the flux density is very much larger in the permeable shells than that in the neighbouring free space is well demonstrated in the figures.

The importance of the deflection of magnetic lines of force in achieving a high shielding effect can also be demonstrated by a three-shell model in which the free spaces between the shells are filled with magnetic matter having a permeability of 500 as shown in Fig. 2.28. The permeability of the shells is taken as 1000. For such a model, the distribution of magnetic lines of force is obtained as shown in Fig. 2.29. The magnetic flux density in the $z=0$ plane is also obtained as shown in Fig. 2.30. We see in Fig. 2.29 that the deflection of the magnetic lines of force is not as marked in the outer four shells in this case. The contrast

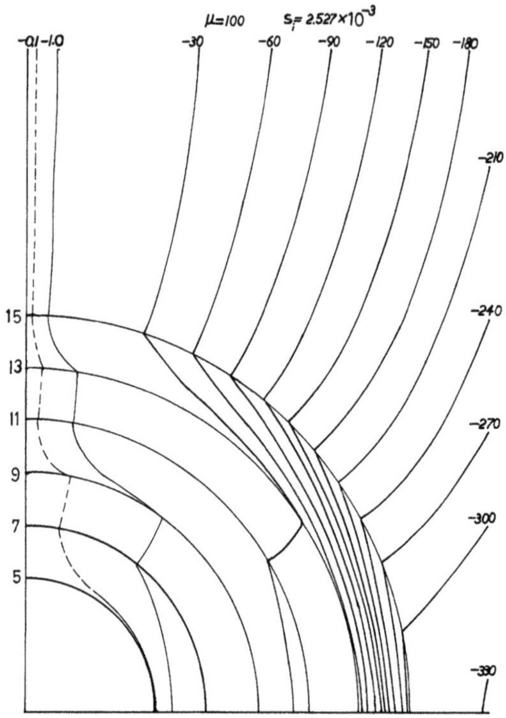

FIG. 2.24 Magnetic lines of force for a 3-shell model of $\mu=100$.

between the flux densities is also not large as can be seen in Fig. 2.30.

The value of S_i for this case becomes larger than that for the models involving free space between the magnetic shells by a factor of 1000 or so.

2.2.2.4 Optimal model of multiple shells for shielding

Let us take a three-shell model such as shown in Fig. 2.31 in which it is assumed that the thickness of the shells and the free space sandwiched between the shells take constant values x and y, respectively. The outermost and innermost radii of the model are respectively given by $R_1=15$ and $R_6=5$, so that we have

$$\begin{aligned} R_1&=15\\ R_2&=15-x\\ R_3&=10+x/2\\ R_4&=10-x/2\\ R_5&=5+x\\ R_6&=5. \end{aligned} \tag{2.65}$$

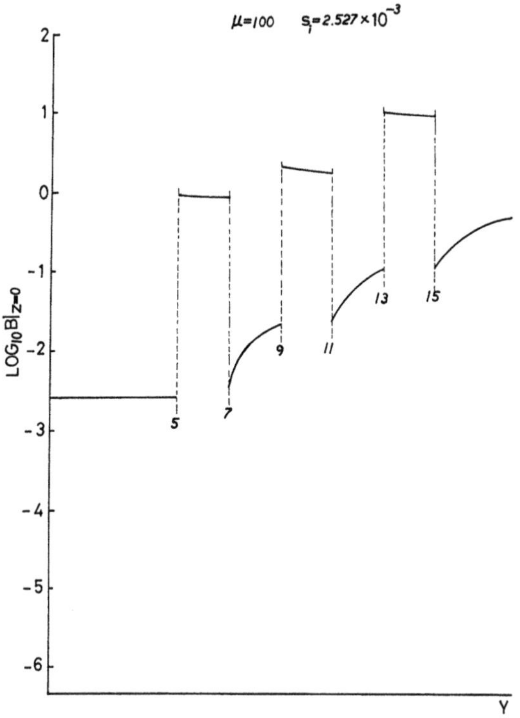

FIG. 2.25 Logarithmic z component of magnetic flux density in the $z=0$ plane for the model shown in Fig. 2.24.

We also have

$$3x+2y=10$$

or

$$y=5-\frac{3}{2}x \quad \left(0<x<\frac{10}{3}\right). \tag{2.66}$$

Going back to Eq. (2.59) and the associated mathematical relations, we can calculate S_i for various values of x ranging from 0 to 10/3, while the permeability μ is taken as a parameter. In the actual calculations, $\mu=100$ and 1000 are assumed. Changes in S_i with x are illustrated in Fig. 2.32.

It is interesting to note that S_i takes on its minimum value at $x=2.01$ and 1.98 respectively for $\mu=100$ and 1000. We therefore see that the highest shielding effect can be obtained by a model in which x is approximately equal to 2.

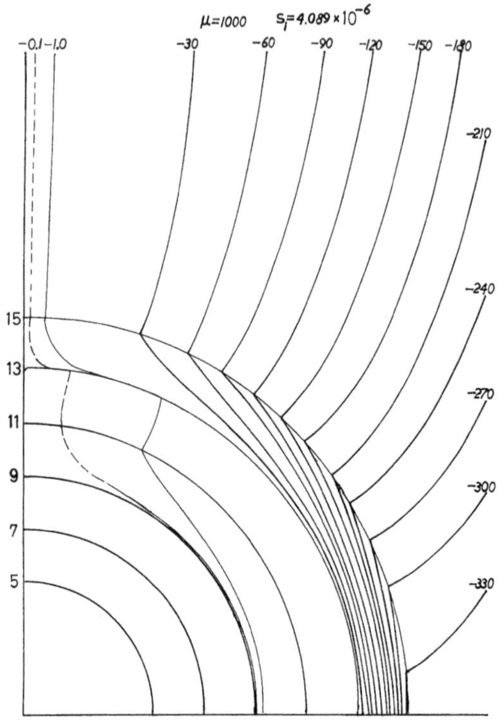

FIG. 2.26 Magnetic lines of force for a 3-shell model of $\mu=1000$.

One may simple-mindedly presume that the thicker the shell is, the higher is the shielding of the model. That it is not so is clearly manifested in Fig. 2.32. Such an effect is caused by the coupling between the outer and inner shells.

2.3 *Magnetic Shielding by Cylindrical Shells*

In the case of experimental studies of rock magnetism, it often becomes necessary to shield rock samples from external magnetic fields in order to avoid possible contamination. In such cases, the samples are usually placed in a single or in multiple cylindrical shells made of highly permeable material.

To the author's knowledge, however, no quantitative study on magnetic shielding by cylindrical shells has so far been conducted. Nothing has been discussed about the shielding effect when the cylindrical shells are finite in length. So far, only cylindrical shells of

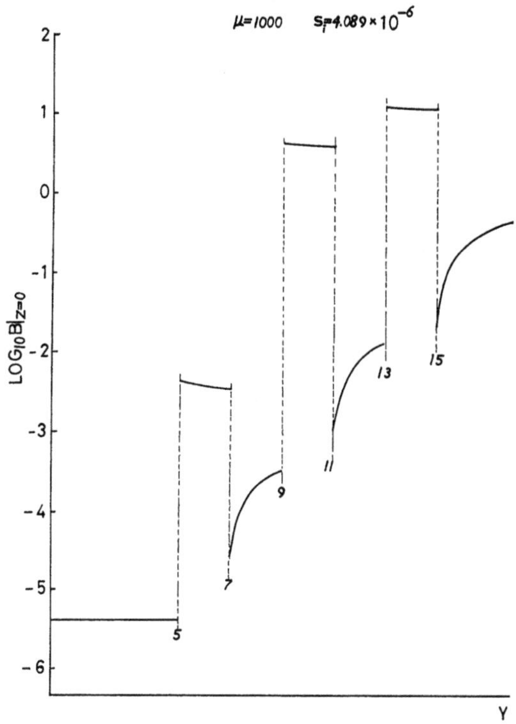

FIG. 2.27 Logarithmic z component of magnetic flux density in the $z=0$ plane for the model shown in Fig. 2.26.

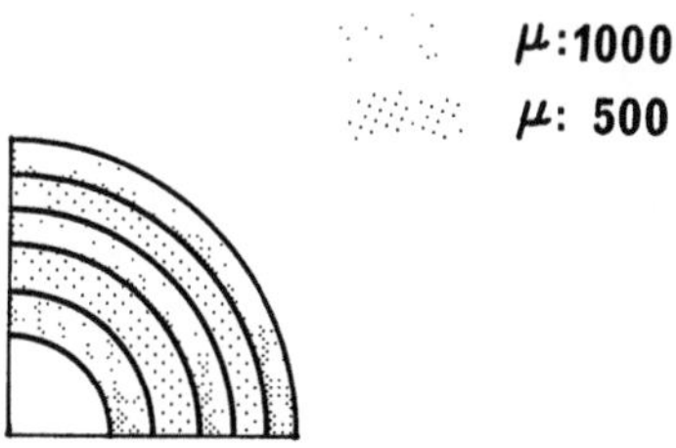

FIG. 2.28 Five-shell model alternatively layered by shells of $\mu=1000$ and $\mu=500$.

infinite length have been studied. Under these circumstances, this section is mainly devoted to determining the degree of shielding by single and multiple cylindrical shells of finite length in an external magnetic field either parallel or perpendicular to the cylinder axis.

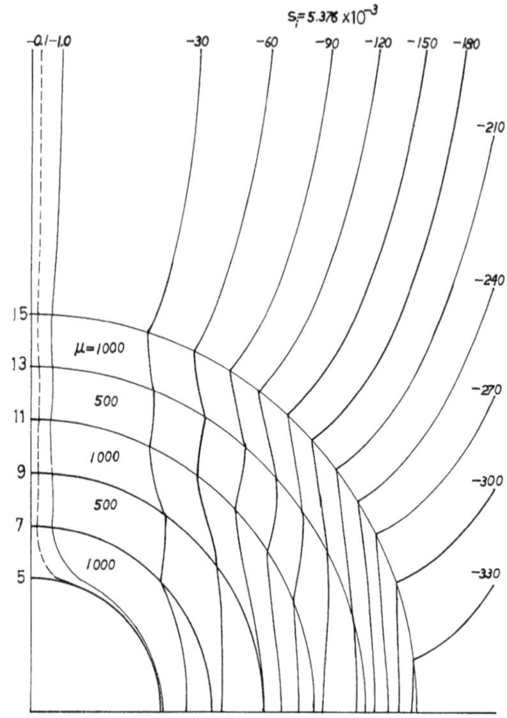

FIG. 2.29 Magnetic lines of force for the model shown in Fig. 2.28.

2.3.1 Cylindrical shell of infinite length

Let us take a hollow cylinder of which the outer and inner radii are denoted respectively by r_1 and r_2. The z axis of cylindrical coordinates (r, θ, z) coincides with the central axis of the cylindrical shell which is infinitely long. At locations very far from the axis, we assume a uniform magnetic field H_0 which is parallel to the $\theta=0$ direction. The permeability of the cylindrical shell is taken as μ.

The magnetic potentials for regions $r>r_1$, $r_1>r>r_2$, and $r_2>r$ are given as

$$\begin{aligned} W_0 &= \left(-H_0 r + \frac{I_0}{r}\right)\cos\theta && r>r_1 \\ W_1 &= \left(-H_1 r + \frac{I_1}{r}\right)\cos\theta && r_1>r>r_2 \qquad (2.67) \\ W_2 &= -H_2 r\cos\theta && r_2>r, \end{aligned}$$

respectively.

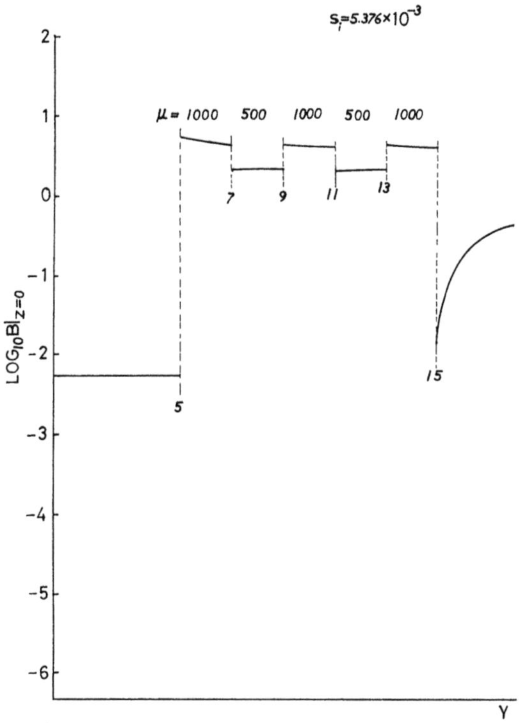

FIG. 2.30 Logarithmic z component of magnetic flux density in the $z=0$ plane for the model shown in Fig. 2.28.

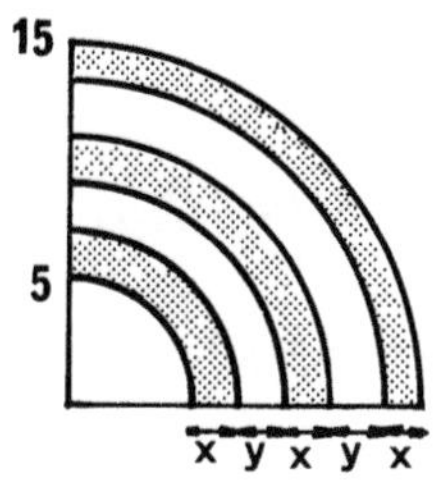

FIG. 2.31 A 3-shell model. The thickness of the shells and the free space sandwiched between them take constant values, x and y, respectively.

The boundary conditions are similar to (2.8). They are

$$
\begin{aligned}
W_0 &= W_1, \quad \frac{\partial W_0}{\partial r} = \mu \frac{\partial W_1}{\partial r} \qquad \text{at } r = r_1 \\
W_1 &= W_2, \quad \mu \frac{\partial W_1}{\partial r} = \frac{\partial W_2}{\partial r} \qquad \text{at } r = r_2
\end{aligned}
\tag{2.68}
$$

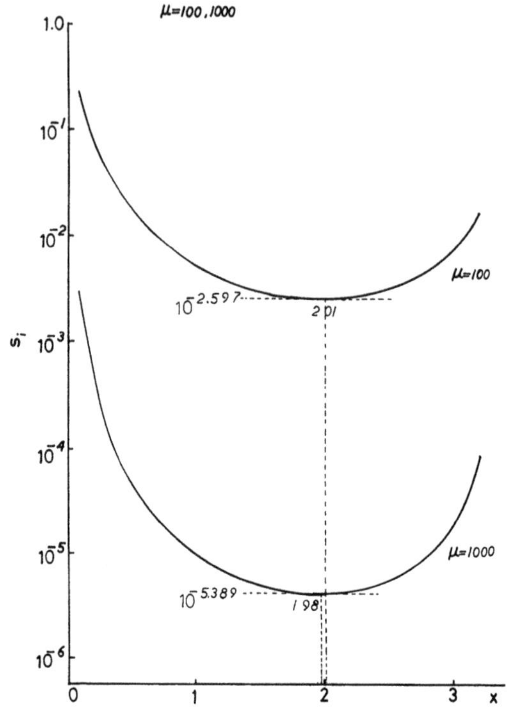

FIG. 2.32 Shielding ratio S_i as a function of magnetic shell thickness for the 3-shell model shown in Fig. 2.31. Two cases $\mu=100$ and 1000 are considered.

which lead to

$$-H_0 r_1+\frac{I_0}{r_1}=-H_1 r_1+\frac{I_1}{r_1}$$
$$-H_0-\frac{I_0}{r_1^2}=\mu\left(-H_1-\frac{2I_1}{r_1^2}\right) \tag{2.69}$$

$$-H_1 r_2+\frac{I_1}{r_2}=-H_2 r_2$$
$$\mu\left(-H_1-\frac{I_1}{r_2^2}\right)=-H_2 . \tag{2.70}$$

Solving (2.69) and (2.70), we obtain

$$I_0=\frac{r_1^2}{4}\frac{\mu^2-1}{\mu}\left(1-\frac{r_2^2}{r_1^2}\right)H_2$$
$$H_1=\frac{\mu+1}{2\mu}H_2 \tag{2.71}$$

$$I_1 = -\frac{\mu-1}{2\mu} r_2^2 H_2$$

$$H_2 = \frac{H_0}{1+\frac{(\mu-1)^2}{4\mu}\left(1-\frac{r_2^2}{r_1^2}\right)}$$

so that the shielding ratio or the ratio of the intensity of internal to external magnetic fields becomes

$$S_i = \frac{1}{1+\frac{(\mu-1)^2}{4\mu}\left(1-\frac{r_2^2}{r_1^2}\right)}^{*)} . \tag{2.72}$$

When $r_2/r_1 \to 0$, S_i tends to

$$S_{i0} = \frac{1}{1+\frac{(\mu-1)^2}{4\mu}} = \frac{4\mu}{(\mu+1)^2} . \tag{2.73}$$

In a fashion similar to the shielding by a spherical shell as given in (2.18), we see that the shielding ratio for a cylindrical shell cannot become smaller than S_{i0} given in (2.73). S_i's for various μ and r_2/r_1 are shown in Table 2.2. Changes in S_i for $r_2/r_1=0.9$ and 0.8 with μ are shown in Fig. 2.33 in which changes in S_{i0} are also shown. Comparing

TABLE 2.2 Magnetic shielding ratio S_i for a cylindrical shell of infinite length.

r_2/r_1 μ	0.9	0.8	0.4	0.0
1	1.00000	1.00000	1.00000	1.00000
5	0.86806	0.77640	0.59809	0.55556
10	0.72215	0.57837	0.37023	0.33058
50	0.30479	0.18791	0.09022	0.07689
100	0.17682	0.10182	0.04633	0.03921
500	0.04056	0.02182	0.00947	0.00797
1000	0.02066	0.01101	0.00475	0.00399
5000	0.00419	0.00222	0.00095	0.00080

*) S_i for a magnetic field proportional to $\cos n\theta$ or $\sin n\theta$ becomes

$$S_i = \frac{1}{1+\frac{(\mu-1)^2}{4\mu}\left(1-\frac{r_2^{2n}}{r_1^{2n}}\right)}$$

as will be given in (3.132).

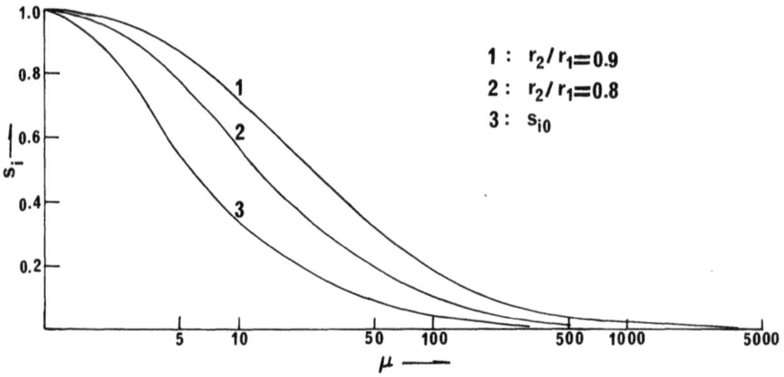

FIG. 2.33 Changes in shielding ratio S_i for a cylindrical shell of infinite length as magnetic permeability increases. The curves 1 and 2 are respectively for shells for which the ratio of outer radius r_1 to inner radius r_2 amounts to 0.9 and 0.8. S_{i0} as shown by curve 3 gives the changes for the limiting case $r_2/r_1 \to 0$.

Fig. 2.33 to Fig. 2.2, it is seen that the degree of shielding by a cylindrical shell is not much different from that by a spherical shell as long as the ratio of inner to outer radii is the same.

The theory of shielding by multiple cylindrical shells can readily be advanced in a manner similar to the case of multiple spherical shells. Since we have seen that the shielding by a cylindrical shell of infinite length is more or less the same as that by a spherical shell, there is good reason to believe that the shielding by multiple cylindrical shells does not differ much from that by multiple spherical shells. It is thus surmised that what we learned about multiple spherical shells approximately holds good for multiple cylindrical shells.

No calculation have been made in this section of the distribution of magnetic lines of force around single or multiple cylindrical shells of infinite length. As can be seen in Fig. 2.5 for a circular cylinder having an infinitely high permeability, the distribution pattern of magnetic lines of force around a cylindrical shell of infinite length is much the same as that in a cross section of a spherical shell.

2.3.2 Cylindrical shell of finite length

As mentioned earlier in this section, it is important for practical purposes to estimate the shielding effect of a cylindrical shell of finite length. Unfortunately, however, no analytical approach to solving such a problem can be found because of the complex boundary conditions.

It is planned in this subsection to solve the problem numerically by making use of the relaxation method for axisymmetric cases. It will be shown in this subsection that the numerical method can readily be applied to cases for multiple cylindrical shells and for a shell with a lid at the ends of the cylinder.

2.3.2.1 External magnetic field parallel to the cylinder axis

The problem of shielding by cylindrical shells of finite length placed in an external magnetic field parallel to the cylinder axis can be tackled by solving differential equations either for magnetic potential W or magnetic flux function ψ. In order to obtain the distribution pattern of magnetic lines of force, we choose to solve the equation for ψ which is given as

$$\frac{\partial}{\partial r}\left(\frac{1}{r}\frac{\partial\psi}{\partial r}\right)+\frac{\partial}{\partial z}\left(\frac{1}{r}\frac{\partial\psi}{\partial z}\right)=0 \tag{2.74}$$

where r and z refer to cylindrical coordinates (r, θ, z), the z axis coinciding with the cylinder axis. Equation (2.74) has already been given in (2.30). It is obvious that the problem is axisymmetric.

Difference equation

As shown in Fig. 2.34, the rz plane is covered by square nets. Taking an arbitrary grid point denoted by 0, the grid points around 0

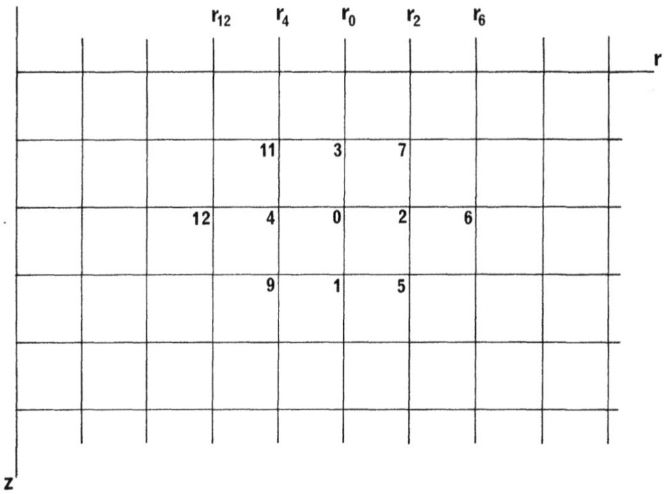

FIG. 2.34 Relaxation net in the rz plane. The distance between adjacent grid points is d.

are numbered as 1, 2, 3, 4, ····· as shown in the figure. The grid interval is d.

Assuming that d is small enough, we can approximately obtain the difference forms of the derivatives at grid point 0 as

$$\begin{aligned}
\left(\frac{\partial\psi}{\partial r}\right)_0 &= \frac{\psi_2-\psi_4}{2d} \\
\left(\frac{\partial^2\psi}{\partial r^2}\right)_0 &= \frac{\psi_2+\psi_4-2\psi_0}{d^2} \\
\left(\frac{\partial\psi}{\partial z}\right)_0 &= \frac{\psi_3-\psi_1}{2d} \\
\left(\frac{\partial^2\psi}{\partial z^2}\right)_0 &= \frac{\psi_1+\psi_3-2\psi_0}{d^2}.
\end{aligned} \tag{2.75}$$

At grid point 0, Eq. (2.74) can be rewritten as

$$\left(\frac{\partial^2\psi}{\partial r^2}\right)_0 - \frac{1}{r_0}\left(\frac{\partial\psi}{\partial r}\right)_0 + \left(\frac{\partial^2\psi}{\partial z^2}\right)_0 = 0. \tag{2.76}$$

Substituting (2.75) into (2.76), we obtain a difference equation as

$$\psi_1+\left(1-\frac{d}{2r_0}\right)\psi_2+\psi_3+\left(1+\frac{d}{2r_0}\right)\psi_4-4\psi_0=0. \tag{2.77}$$

Similarly, we obtain

$$\psi_5+\left(1-\frac{d}{2r_2}\right)\psi_6+\psi_7+\left(1+\frac{d}{2r_2}\right)\psi_0-4\psi_2=0 \tag{2.78}$$

for grid point 2.

Since equations such as (2.77) and (2.78) can be obtained for all the grid points, solution of the problem requires finding ψ_0, ψ_1, ψ_2, ··· which satisfy these equations simultaneously in addition to satisfying the proper boundary conditions. In actual calculations, we rely on the relaxation method (e.g., SOUTHWELL, 1946, 1956) details of which are not described here.

Boundary conditions

The components of magnetic flux density are given by

$$B_r = -\frac{1}{r}\frac{\partial\psi}{\partial z}, \quad B_z = \frac{1}{r}\frac{\partial\psi}{\partial r}. \tag{2.79}$$

Using these relations, the continuity conditions, on the tangential component of magnetic field and on the normal component of magnetic flux density, can be described in terms of ψ.

(a) Boundary perpendicular to the z axis

Let us consider a boundary shown in Fig. 2.35. ψ and ψ' denote the magnetic flux function for media I and II, respectively. μ_1 and μ_2 are the magnetic permeabilities for the respective media. From the continuity of the normal component of magnetic flux density at the boundary, we obtain

$$\frac{1}{r}\frac{\partial\psi}{\partial r}=\frac{1}{r}\frac{\partial\psi'}{\partial r}$$

there. On integrating both sides of the above equation with respect to r and putting the integration constant to zero, we obtain

$$\psi=\psi'$$

so that we have

$$\begin{aligned}\psi_2&=\psi_2'\\ \psi_0&=\psi_0'\\ \psi_4&=\psi_4'\end{aligned} \tag{2.80}$$

at the boundary.

The continuity of the tangential component of magnetic field leads to

$$\frac{1}{\mu_1}\frac{\partial\psi}{\partial z}=\frac{1}{\mu_2}\frac{\partial\psi'}{\partial z}$$

at the boundary. The difference form of the equation is

$$\frac{1}{\mu_1}\frac{\psi_3-\psi_1}{2d}=\frac{1}{\mu_2}\frac{\psi_3'-\psi_1'}{2d}$$

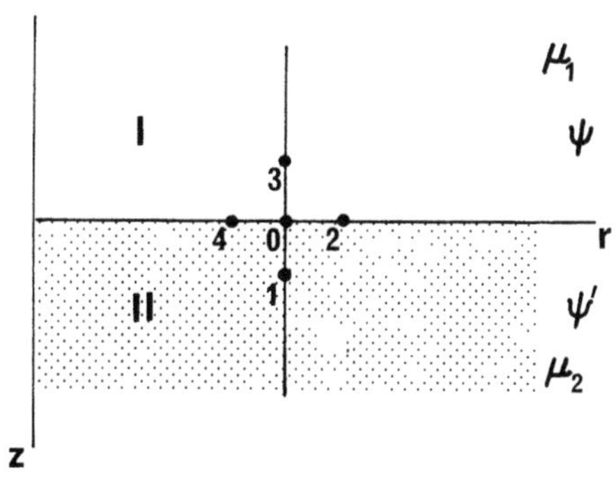

FIG. 2.35 Grid points for describing boundary conditions.

which can be rewritten

$$T(\psi_3-\bar{\psi}_1)=\bar{\psi}_3'-\psi_1' \tag{2.81}$$

where $\bar{\psi}$ and $\bar{\psi}'$ are the hypothetical values of ψ and ψ' when media I and II are assumed to exist beyond the boundary. In (2.81), T is defined by

$$T=\mu_1/\mu_2\,. \tag{2.82}$$

In such hypothetical media, Eq. (2.74) and associated equations should hold good, so that we have

$$\bar{\psi}_1+A\psi_2+\psi_3+B\psi_4-4\psi_0=0 \tag{2.83}$$

and

$$\psi_1'+A\psi_2'+\bar{\psi}_3'+B\psi_4'-4\psi_0'=0 \tag{2.84}$$

where

$$\begin{aligned} A&=1-\frac{d}{2r_0} \\ B&=1+\frac{d}{2r_0}\,. \end{aligned} \tag{2.85}$$

From (2.83), we obtain

$$\bar{\psi}_1=4\psi_0-A\psi_2-\psi_3-B\psi_4\,. \tag{2.86}$$

From (2.84) and (2.80), we obtain

$$\bar{\psi}_3'=4\psi_0-\psi_1'-A\psi_2-B\psi_4\,. \tag{2.87}$$

Substituting (2.86) and (2.87) into (2.81), we obtain

$$C\psi_1'+A\psi_2+D\psi_3+B\psi_4-4\psi_0=0 \tag{2.88}$$

where

$$\begin{aligned} C&=\frac{2}{1+T} \\ D&=\frac{2T}{1+T}\,. \end{aligned} \tag{2.89}$$

We, therefore, see that (2.88) is the difference equation to be adopted at the boundary.

When the configuration of magnetic matter is as shown in Fig. 2.36, the difference equation to be satisfied at the boundary becomes

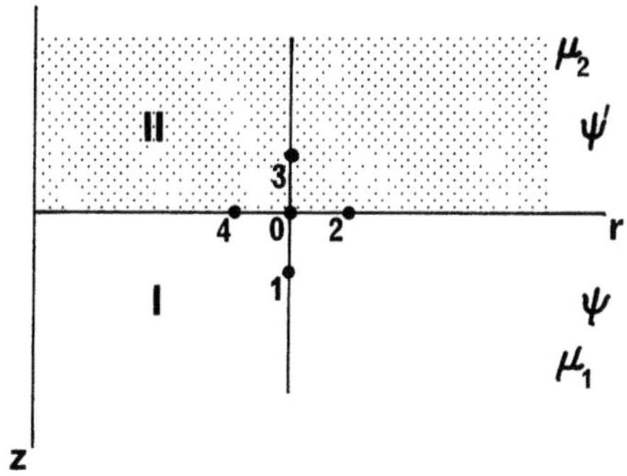

FIG. 2.36 Grid points for describing boundary conditions.

$$D\psi_1+A\psi_2+C\psi_3'+B\psi_4-4\psi_0=0 \tag{2.90}$$

which may be derived in more or less the same way as (2.88).

(b) Boundary perpendicular to the r axis

Let us now consider a boundary which is shown in Fig. 2.37. From the continuity of the normal component of magnetic flux density, we obtain

$$\frac{1}{r}\frac{\partial\psi}{\partial z}=\frac{1}{r}\frac{\partial\psi'}{\partial z}$$

at the boundary, so that

$$\begin{aligned}\psi_1&=\psi_1'\\ \psi_3&=\psi_3'\\ \psi_0&=\psi_0'.\end{aligned} \tag{2.91}$$

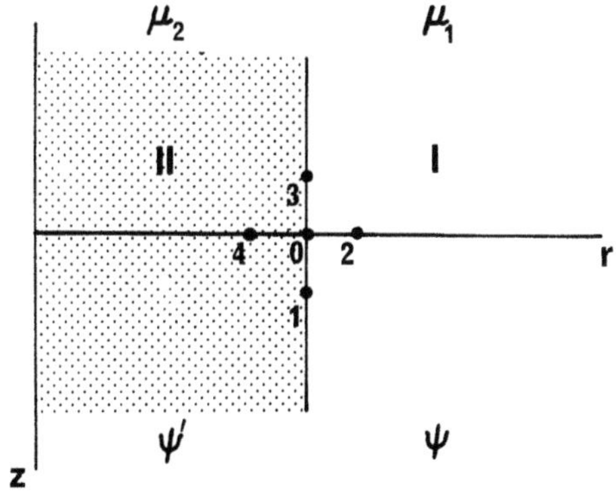

FIG. 2.37 Grid points for describing boundary conditions.

Continuity of the tangential component of magnetic field leads to

$$\frac{1}{\mu_1}\frac{1}{r}\frac{\partial\psi}{\partial r}=\frac{1}{\mu_2}\frac{1}{r}\frac{\partial\psi'}{\partial r}$$

of which the difference form is

$$\frac{1}{\mu_1}\frac{1}{2d}(\psi_2-\bar{\psi}_4)=\frac{1}{\mu_2}\frac{1}{2d}(\bar{\psi}_2'-\psi_4')$$

which may be further rewritten as

$$T(\psi_2-\bar{\psi}_4)=\bar{\psi}_2'-\psi_4' \tag{2.92}$$

taking (2.82) into account. $\bar{\psi}_4$ and $\bar{\psi}_2'$ in the above expressions are again the hypothetical values in the respective regions beyond the boundary.

As we know that

$$\psi_1+A\psi_2+\psi_3+B\bar{\psi}_4-4\psi_0=0 \tag{2.93}$$

and

$$\psi_1'+A\bar{\psi}_2'+\psi_3'+B\psi_4'-4\psi_0'=0 \tag{2.94}$$

hold good respectively in media I and II, we obtain

$$\bar{\psi}_4=\frac{1}{B}(4\psi_0-\psi_1-A\psi_2-\psi_3) \tag{2.95}$$

and

$$\bar{\psi}_2'=\frac{1}{A}(4\psi_0-\psi_1-\psi_3-B\psi_4') \tag{2.96}$$

with the aid of (2.91).

Eliminating $\bar{\psi}_4$ and $\bar{\psi}_2'$ from (2.92) by substituting (2.95) and (2.96) into (2.92), we obtain

$$\psi_1+E\psi_2+\psi_3+F\psi_4'-4\psi_0=0 \tag{2.97}$$

where

$$E=T(1+A/B)/(1/A+T/B)$$

and

$$F=(1+B/A)/(1/A+T/B). \tag{2.98}$$

(2.97) is the difference equation which must be satisfied at the boundary.

When medium I is surrounded by medium II as shown in Fig. 2.38, we readily see that the difference equation at the boundary is

$$\psi_1+G\psi_2'+\psi_3+H\psi_4-4\psi_0=0 \tag{2.99}$$

where

$$G=(1+A/B)/(T/A+1/B)$$

and

$$H=T(1+B/A)/(T/A+1/B). \tag{2.100}$$

In the present problem, it is assumed that a uniform magnetic field H_0 parallel to the z axis exists at a great distance from the cylindrical shell. The magnetic flux function is therefore given as

$$\psi=\frac{1}{2}r^2H_0 \tag{2.101}$$

there. This provides the boundary condition at a great distance.

Relaxation

In order to solve difference equations such as (2.77) and (2.78), we take residuals defined by

$$R_1=\psi_1+\left(1-\frac{d}{2r_0}\right)\psi_2+\psi_3+\left(1+\frac{d}{2r_0}\right)r_4-4\psi_0$$

$$R_2=\psi_5+\left(1-\frac{d}{2r_2}\right)\psi_6+\psi_7+\left(1+\frac{d}{2r_2}\right)r_0-4\psi_2$$

......

......

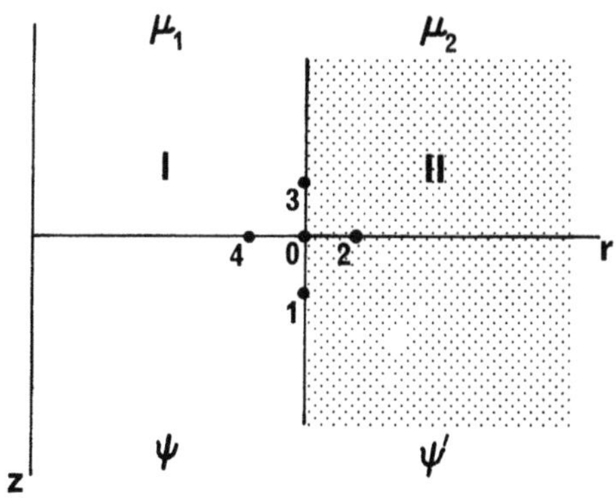

FIG. 2.38 Grid points for describing boundary conditions.

at respective grid points. These residuals are then made zero step by step. Of course, the boundary conditions in the above must also be satisfied. No details are given of the relaxation method used here because it follows more or less the standard approach of relaxation calculus.

Because of symmetry, it is necessary to perform computations only over a quarter of the space. Grid points amounting to 31×50 are used for the actual calculation. It is clear that ψ is zero on the z axis from the symmetry consideration.

Test solutions

In order to test the computer program for the relaxation method used, the distribution of magnetic lines of force around a thin circular disk placed at $z=0$ perpendicular to the external magnetic field is examined. It is assumed that the normal component of magnetic field vanishes at the disk, so that the model is equivalent to electromagnetic induction in a perfectly conducting disk. It is also seen that the magnetic lines of force represent the stream lines of an ideal fluid flow passing the disk.

The condition that the normal magnetic field vanishes leads to $\psi=$const. at the disk. As $\psi=0$ should hold good on the z axis, $\psi=0$ holds good at the disk. Figure 2.39 shows the distribution of magnetic lines of force obtained by performing the relaxation computation. It is clearly seen in the figure that the magnetic lines of force are deflected by the conductor. The result agrees with that of LAMB (1932), who solved the flow of an ideal fluid around a circular disk using spheroidal coordinates, and with that of RIKITAKE (1966) who applied the relaxation method manually to the problem. Judging from the good agreement between them, the relaxation program adopted should be applicable to more complicated problems.

Looking at Fig. 2.39, we notice that the interval between neighbouring lines of force becomes smaller as r increases. This is inevitable because of the axisymmetric treatment as mentioned in the last section.

In addition to Fig. 2.39, Fig. 2.40 provides another example showing the magnetic lines of force when a disk is added to the former case. It should be noted that there are actually three disks, but only a quarter of the system is shown in the figure because of symmetry consideration.

Single cylindrical shell

Once the computer program for the relaxation has been

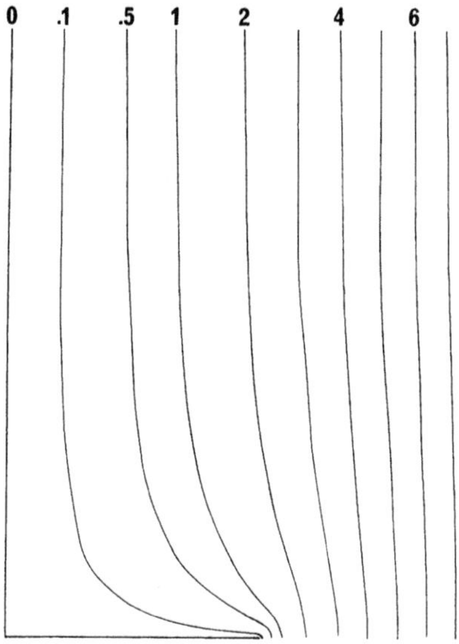

FIG. 2.39 Magnetic lines of force around a thin circular disk when the normal component of magnetic field vanishes at the disk.

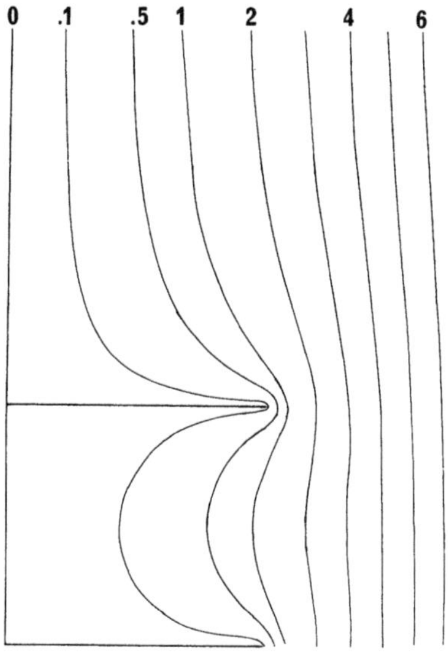

FIG. 2.40 Magnetic lines of force around a set of thin circular disks when the normal component of magnetic field vanishes at the disks. Only one-quarter of the space is shown in the figure.

completed, it is easy to study the distribution patterns of magnetic lines of force around a cylindrical shell having any thickness. As typical examples, Figures 2.41, 2.42 and 2.43 show the three cases for which the combinations of thickness and permeability are (2*d*, 100), (2*d*, 1000) and (4*d*, 100), respectively. *d* is the grid interval for the numerical calculation. The length and the outer radius of the shell are 30*d* and 10*d*, respectively. The permeability in free space is taken to be unity as before.

Figures 2.41, 2.42 and 2.43 clearly demonstrate that the magnetic lines of force are attracted by the highly permeable shell, so that the magnetic flux density inside the shell becomes low.

Let us define the shielding ratio by

$$S_i = B_z / B_{z0} \tag{2.102}$$

in this case, where B_z is the magnetic flux density along the z axis. B_{z0} is the value of B_z for the external field in the absence of the shell.

B_z is defined by

$$B_z = \lim_{r \to 0} \frac{1}{r} \frac{\partial \psi}{\partial r} .$$

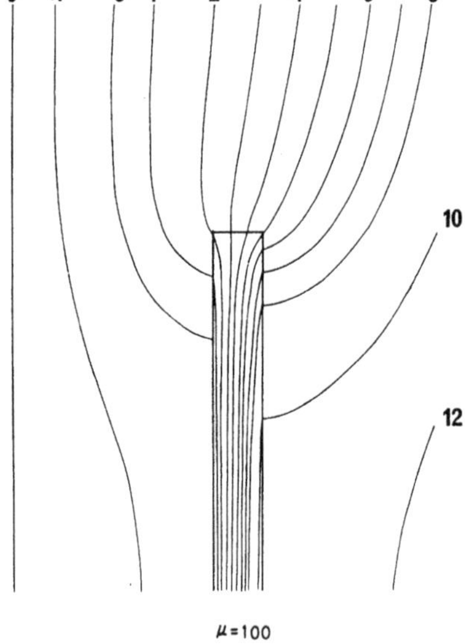

FIG. 2.41 Magnetic lines of force around a cylindrical shell (length: 30*d*, outer radius: 10*d*, thickness: 2*d*, $\mu = 100$).

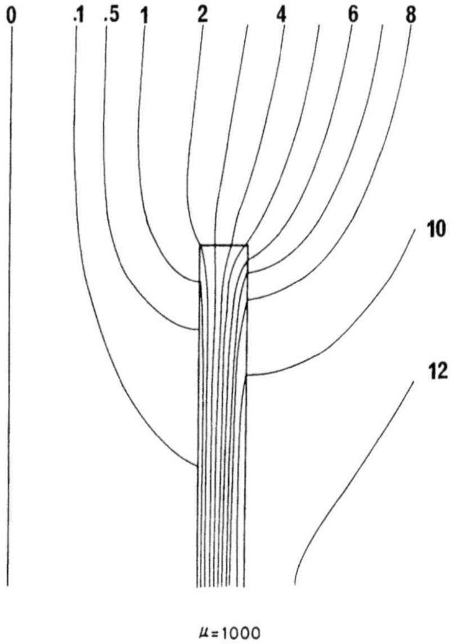

FIG. 2.42 Magnetic lines of force around a cylindrical shell (length: 30d, outer radius: 10d, thickness: 2d, μ=1000).

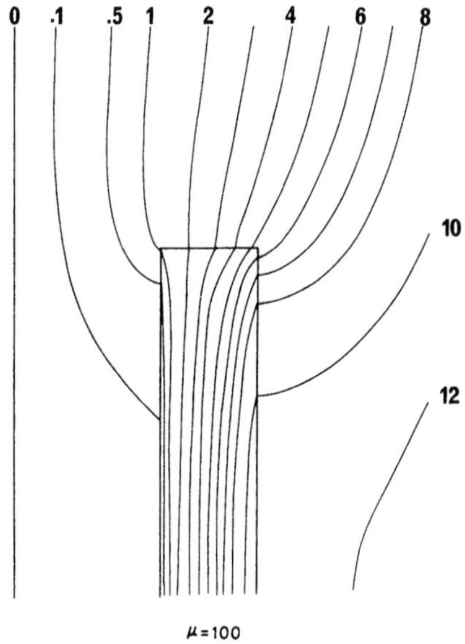

FIG. 2.43 Magnetic lines of force around a cylindrical shell (length: 30d, outer radius: 10d, thickness: 4d, μ=100).

Because of the axisymmetry, we take

$$\lim_{r\to 0}\frac{\partial\psi}{\partial r}=0$$

so that

$$B_z=\frac{\partial^2\psi}{\partial r^2}$$

which can be written in a difference form such as

$$B_z=\frac{\psi_2+\psi_4-2\psi_0}{d^2}$$

referring to Fig. 2.34.

As grid point 0 is taken on the z axis in the present example, we have

$$\psi_2=\psi_4$$
$$\psi_0=0$$

so that finally

$$B_z=\frac{2\psi_2}{d^2}\,. \tag{2.103}$$

In addition, (2.101) leads to

$$B_{z0}=0.5H_0\,. \tag{2.104}$$

The shielding ratio S_i as defined by (2.102) can then be calculated from (2.103) and (2.104). Figures 2.44 and 2.45 show how S_i changes as z increases for the shells having thicknesses $2d$ and $4d$, respectively. The points represented by circles, squares and triangles in the figures indicate that they are for shells of $\mu=100$, 500 and 1000, respectively. Tables such as 2.1 and 2.2, in which S_i was given for various combinations of μ and shell thickness, are not provided here because of low numerical accuracy.

In Figs. 2.44 and 2.45, the plots are not quite smooth, due to inaccuracies in the numerical calculations. Despite such inaccuracies, we see that cylindrical shells parallel to the external magnetic field can shield the inside of the shell to some extent. With various combinations of shell geometry and magnetic permeability, it is possible to reduce the magnetic field by a factor of ten or more.

It is also observed in Figs. 2.44 and 2.45 that the higher the permeability and the thicker the shell, the larger is the degree of

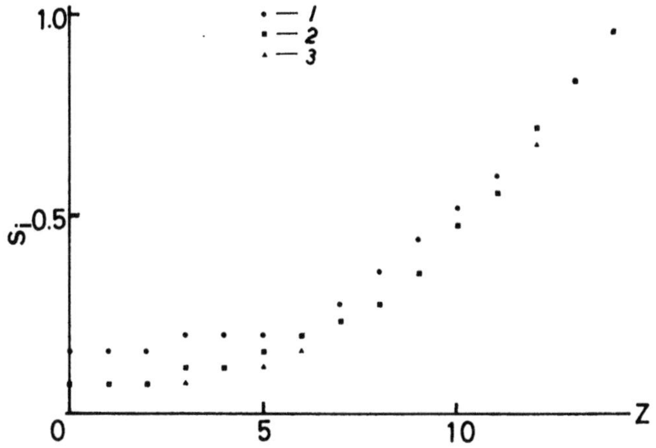

FIG. 2.44 Changes in shielding ratio S_i along the z axis for a cylindrical shell of thickness, length and outer radius, $2d$, $30d$ and $10d$, respectively. Plots by circles (1), squares (2) and triangles (3) respectively indicate values for $\mu=100$, 500 and 1000.

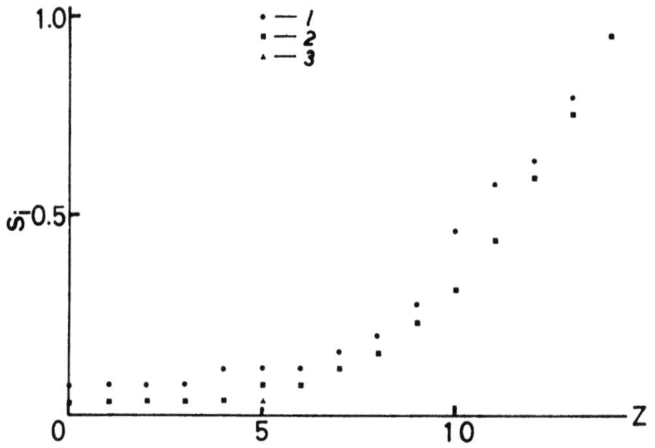

FIG. 2.45 Changes in shielding ratio S_i along the z axis for a cylindrical shell of thickness, length and outer radius, $4d$, $30d$ and $10d$, respectively. Plots by circles (1), squares (2) and triangles (3) respectively indicate values for $\mu=100$, 500 and 1000.

shielding as expected. The reason why the magnetic flux density near the central axis becomes small is that the magnetic lines of force are attracted by the shell, and so they tend to be deflected from the axis. Such a tendency becomes more marked when the permeability is high and the shell thickness is large.

Multiple cylindrical shells

There is no difficulty in dealing with cases of multiple cylindrical shells. It is interesting to see how the magnetic lines of force are distributed around single, double and triple shells. For that purpose, three cases for which $\mu=100$ and the shell thickness is d are studied as shown in Figs. 2.46, 2.47 and 2.48. As can be seen in the figures, the magnetic lines of force for $\psi=0.1$ is deflected more and more from the central axis as the number of shells increases. We thus see that the degree of shielding increases with the number of shells.

Changes in S_i with z for a single shell, Fig. 2.46, are shown in Fig. 2.49 for $\mu=100$, 500 and 1000. Similar illustrations for the double shells, which are shown in Fig. 2.47, are given in Fig. 2.50. Comparing Fig. 2.49 to Fig. 2.50, we see that the shielding effect is larger for the double shell than that for the single shell if the permeability is the same.

Further calculations indicate that the degree of shielding for a double shell with $\mu=100$ is about the same as that for a single shell of

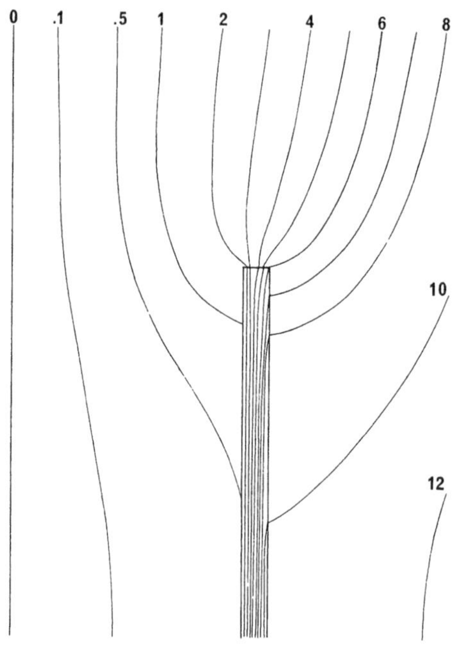

FIG. 2.46 Magnetic lines of force around a single cylindrical shell of thickness d and $\mu=100$.

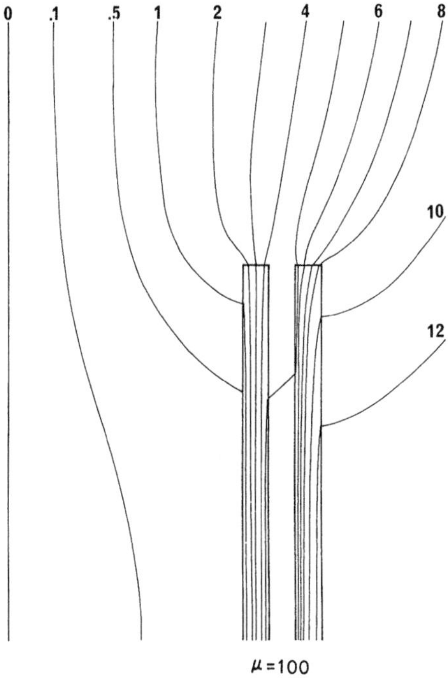

FIG. 2.47 Magnetic lines of force around a double cylindrical shell of thickness d and $\mu=100$.

$\mu=500$. Similarly, triple shells of $\mu=100$ are approximately equivalent to a single shell of $\mu=1000$ as far as shielding power is concerned.

Cylindrical shell with a lid

When a cylindrical shell has a lid at the end, the magnetic lines of force are distributed in the manner shown in Fig. 2.51. Comparing Fig. 2.51 to Fig. 2.46, we see no great difference in the distribution of magnetic lines of force in the space around the center of the shell. This means that the shielding is affected little by the added lid. Actually, the external magnetic lines of force tend to be attracted towards the axis by the lid. Such a tendency plays a negative role as far as shielding is concerned. One might have supposed that a lid would enhance the shielding power. But this is not so as can be seen in Fig. 2.51.

Figure 2.52 shows the magnetic lines of force when there is a hole in the lid. As far as the central space is concerned, no remarkable effect of the hole is observed.

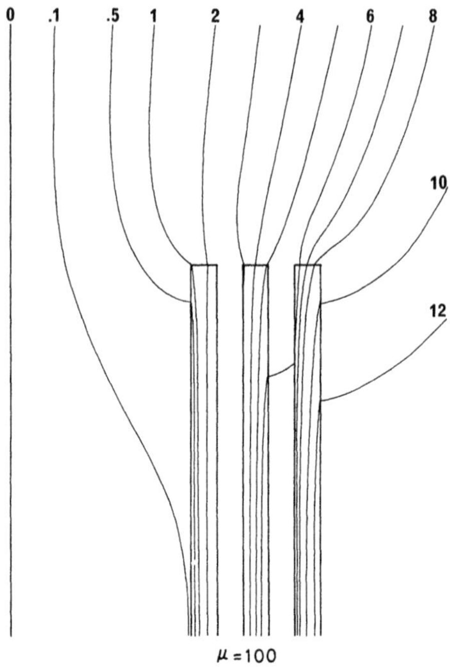

FIG. 2.48 Magnetic lines of force around a triple cylindrical shell of thickness d and $\mu=100$.

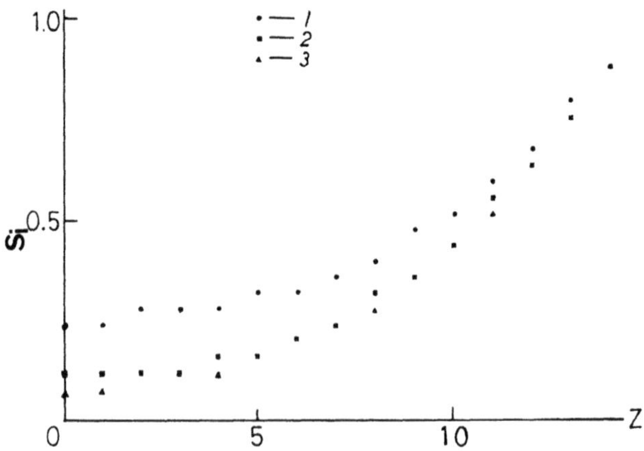

FIG. 2.49 Changes in shielding ratio S_i along the z axis for a single shell shown in Fig. 2.46. Plots by circles (1), squares (2) and triangles (3) respectively indicate values for $\mu=100$, 500 and 1000.

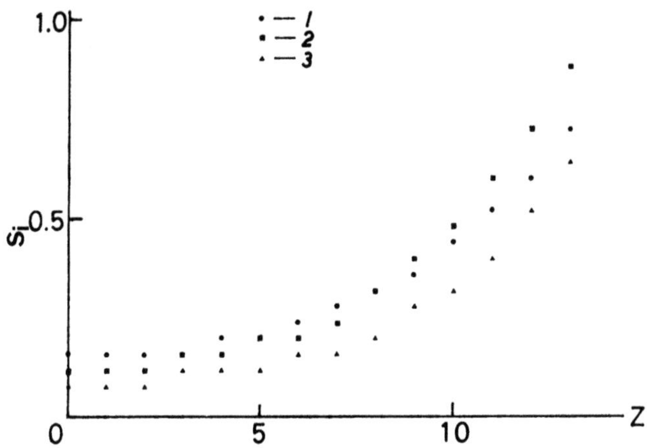

FIG. 2.50 Changes in shielding ratio S_i along the z axis for a double shell shown in Fig. 2.47. Plots by circles (1), squares (2) and triangles (3) respectively indicate values for μ=100, 500 and 1000.

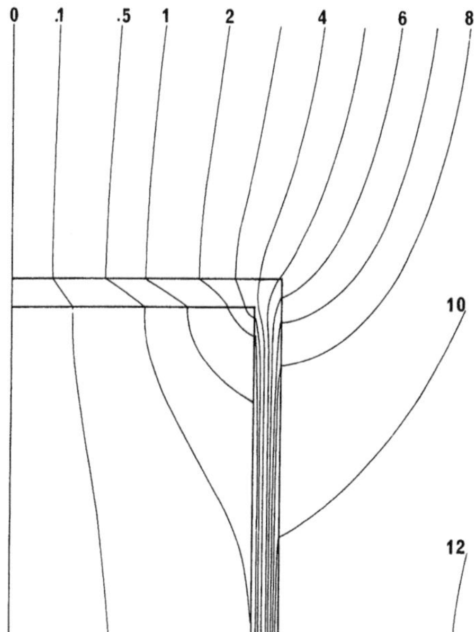

FIG. 2.51 Magnetic lines of force around a cylindrical shell with a lid at its ends.

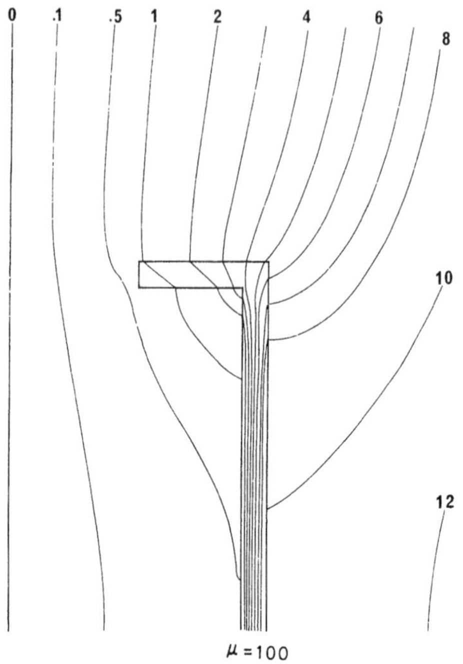

FIG. 2.52 Magnetic lines of force around a cylindrical shell with an imperfect lid at its ends.

In the case of a shell with a lid, it turns out that the magnetic lines of force which enter into the lid from outside are deflected less than those for a spherical shell. The magnetic lines of force are then obliged to leak out from the shell into the internal space, so that only little rarefaction of magnetic lines of force takes place around the center of the shell, resulting in low shielding.

As a demonstration of how the distribution of magnetic lines of force is modified, Figure 2.53 shows the magnetic lines of force in and around a solid magnetic cylinder.

Two-layer shell

In order to examine the shielding effect of a two-layer shell, the distribution of magnetic lines of force around the models shown in Figs. 2.54 and 2.55 are studied. The shells consist of two layers having different permeabilities.

It is apparent in the figures that the magnetic flux density in the space around the center becomes lower for the shell for which the outer

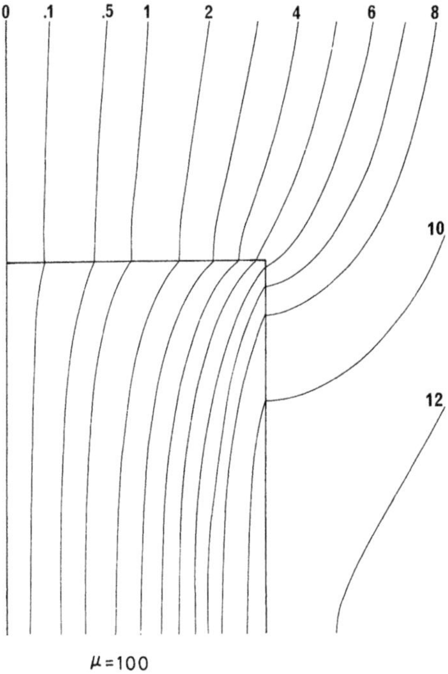

FIG. 2.53 Magnetic lines of force around a solid cylinder.

layer is more permeable than the inner layer, than for the shell for which the outer layer is less permeable than the inner layer.

Figure 2.56 shows the changes in the shielding ratio S_i on the z axis. It is observed that S_i for case 1 ($\mu=1000$ for the inner layer and $\mu=100$ for the outer layer) is larger than that for case 2 ($\mu=100$ for the inner layer and $\mu=1000$ for the outer layer).

2.3.2.2 External magnetic field perpendicular to the cylinder axis

In addition to the previously-treated shielding effect of cylindrical shells of finite length for an external magnetic field parallel to the cylinder axis, it is important to see how effective the shielding would be when the axis of a cylindrical shell of finite length is perpendicular to the external magnetic field.

The problem to be solved here is essentially three-dimensional, so that it is difficult to solve with either a two-dimensional or an axisymmetric approach based on the magnetic flux function as utilized in some of the previous studies.

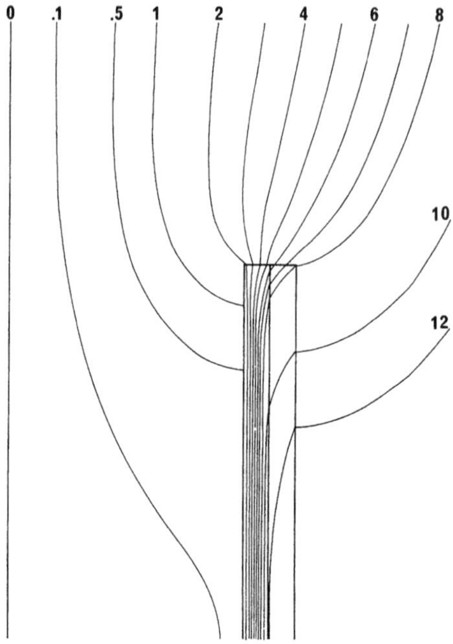

FIG. 2.54 Magnetic lines of force around a two-layer cylindrical shell. The magnetic permeability of the inner shell is higher than that of the outer shell.

Let us then go back to Eq.(2.6) which should be satisfied by the magnetic potential W. Making use of cylindrical coordinates (r, θ, z), (2.6) can be rewritten as

$$\frac{\partial^2 W}{\partial r^2}+\frac{1}{r}\frac{\partial W}{\partial r}+\frac{1}{r^2}\frac{\partial^2 W}{\partial \theta^2}+\frac{\partial^2 W}{\partial z^2}=0. \tag{2.105}$$

Since the external magnetic field, which is directd along the $\theta=0$ axis, is perpendicular to the cylinder axis or the z axis, it is evident that the problem is symmetric about the xz plane, the x axis coinciding with the $\theta=0$ axis. In that case, we can put

$$W=\phi\ (r, z)\cos\theta. \tag{2.106}$$

Substituting (2.106) into (2.105), we obtain

$$\frac{\partial^2 \phi}{\partial r^2}+\frac{1}{r}\frac{\partial \phi}{\partial r}-\frac{\phi}{r^2}+\frac{\partial^2 \phi}{\partial z^2}=0 \tag{2.107}$$

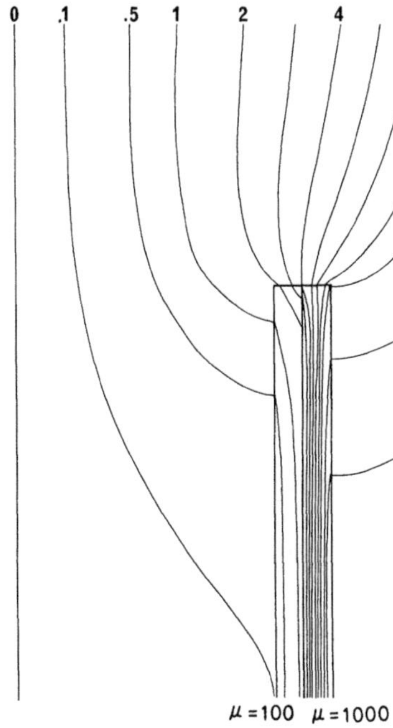

Fig. 2.55 Magnetic lines of force around a two-layer cylindrical shell. The magnetic permeability of the inner shell is lower than that of the outer shell.

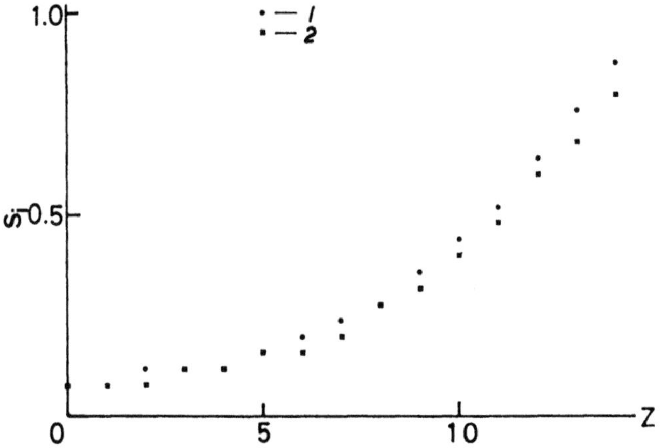

Fig. 2.56 Changes in the shielding ratio S_i along the z axis for two-layer models. Cases 1 and 2 correspond to models shown in Figs. 2.54 and 2.55, respectively.

solving which, with proper boundary conditions determines the modification of the magnetic lines of force of an external uniform magnetic field by a permeable cylindrical shell of finite length. It is difficult in this case, however, to illustrate the distribution of magnetic lines of force because no analytical solutions for ϕ in the xz plane are given. It is not possible in this case to illustrate the distribution of magnetic lines of force on the basis of equations such as (2.21) or (2.23).

Difference equation

Let us cover the rz plane with a net again as shown in Fig. 2.34. With the values of ϕ at grid points 0, 1, 2, 3 and 4, (2.107) at grid point 0 can be written as

$$\phi_1+\left(1-\frac{d}{2r_0}\right)\phi_2+\phi_3+\left(1+\frac{d}{2r_0}\right)\phi_4-\left(\frac{d^2}{r_0^{\,2}}+4\right)\phi_0=0. \qquad (2.108)$$

Similar difference equations can be assigned to each grid pont. These equations can be solved by making use of the relaxation method under appropriate boundary conditions as discussed below.

Boundary conditions

(a) Boundary perpendicular to the z axis

Let us consider a boundary between two media each having permeabilities μ_1 and μ_2 as shown in Fig. 2.35. ψ and ψ' in the figure should read respectively ϕ and ϕ' in the present problem.

As the tangential component of the magnetic field is continuous at the boundary, we put

$$\frac{\partial\phi}{\partial r}=\frac{\partial\phi'}{\partial r}$$

and, therefore, obtain

$$\phi=\phi'$$

without loss of generality. Consequently we have

$$\begin{aligned}\phi_0&=\phi_0'\\ \phi_2&=\phi_2'\\ \phi_4&=\phi_4'.\end{aligned} \qquad (2.109)$$

On the other hand, continuity of the normal component of the magnetic flux density implies

$$\mu_1\frac{\partial\phi}{\partial z}=\mu_2\frac{\partial\phi'}{\partial z}$$

the difference form of which becomes

$$T(\phi_3-\bar{\phi}_1)=\bar{\phi}_3'-\phi_1' \tag{2.110}$$

where $\bar{\phi}_1$ and $\bar{\phi}_1'$ are the hypothetical values of ϕ and ϕ' when media I and II are assumed to exist beyond the boundary. T is defined by (2.82).

In such hypothetical media, Eq. (2.107) and associated equations hold good, so that we have

$$\bar{\phi}_1+A\phi_2+\phi_3+B\phi_4-\left(\frac{d^2}{r_0^2}+4\right)\phi_0=0 \tag{2.111}$$

and

$$\phi_1'+A\phi_2'+\bar{\phi}_3'+B\phi_4'-\left(\frac{d^2}{r_0^2}+4\right)\phi_0'=0 \tag{2.112}$$

where A and B are defined by (2.85).

From (2.111), we obtain

$$\bar{\phi}_1=K\phi_0-A\phi_2-\phi_3-B\phi_4\,. \tag{2.113}$$

From (2.112) and (2.109), we obtain

$$\bar{\phi}_3'=K\phi_0-\phi_1'-A\phi_2-B\phi_4 \tag{2.114}$$

where

$$K=\frac{d^2}{r_0^2}+4\,. \tag{2.115}$$

Substituting (2.113) and (2.114) into (2.110), we obtain

$$C\phi_1'+A\phi_2+D\phi_3+B\phi_4-K\phi_0=0 \tag{2.116}$$

where C and D are defined by (2.89).

Equation (2.116) provides the condition to be satisfied at the boundary.

When the boundary is the one shown in Fig. 2.36, the difference equation to be satisfied at the boundary becomes

$$D\phi_1'+A\phi_2+C\phi_3+B\phi_4-K\phi_0=0\,. \tag{2.117}$$

(b) Boundary perpendicular to the r axis

Let us now consider a boundary which is shown in Fig. 2.37. From the continuity of the tangential component of the magnetic field, we have

$$\frac{\partial \phi}{\partial z}=\frac{\partial \phi'}{\partial z}$$

which leads to

$$\begin{aligned}\phi_1&=\phi_1{}'\\ \phi_3&=\phi_3{}'\\ \phi_0&=\phi_0{}'.\end{aligned}\tag{2.118}$$

On the other hand, continuity of the normal component of the magnetic flux density implies

$$\mu_1\frac{\partial \phi}{\partial r}=\mu_2\frac{\partial \phi'}{\partial r}$$

for which the difference form is

$$T(\phi_2-\bar{\phi}_4)=\bar{\phi}_2{}'-\phi_4{}'\tag{2.119}$$

where $\bar{\phi}_4$ and $\bar{\phi}_2{}'$ are hypothetical values as defined above.

As we know that

$$\phi_1+A\phi_2+\phi_3+B\bar{\phi}_4-K\phi_0=0\tag{2.120}$$

and

$$\phi_1{}'+A\bar{\phi}_2{}'+\phi_3{}'+B\phi_4{}'-K\phi_0=0\tag{2.121}$$

hold good, respectively, in media I and II, we obtain

$$\bar{\phi}_4=\frac{1}{B}(K\phi_0-\phi_1-A\phi_2-\phi_3)\tag{2.122}$$

and

$$\bar{\phi}_2{}'=\frac{1}{A}(K\phi_0-\phi_1-\phi_3-B\phi_4{}')\tag{2.123}$$

with the aid of (2.118).

Substituting (2.122) and (2.123) into (2.119), we obtain

$$\phi_1+E\phi_2+\phi_3+F\phi_4{}'-K\phi_0=0\tag{2.124}$$

where E and F are defined by (2.98).

(2.124) is the difference equation to be satisfied at the boundary.

When the boundary is the one shown in Fig. 2.38, we readily see that the difference equation at the boundary is

$$\phi_1+G\phi_2{}'+\phi_3+H\phi_4-K\phi_0=0\tag{2.125}$$

where G and H are defined by (2.100).

At a great distance from the cylindrical shell, there is a uniform magnetic field parallel to the r direction, so that we may put

$$\phi = -H_0 r. \tag{2.126}$$

This provides the boundary condition at a great distance where the effect of the cylindrical shell is negligible.

Relaxation

The relaxation procedure for the present problem is almost the same as that for the previous case in which the external magnetic field was parallel to the cylinder axis. There is no need, therefore, to repeat the details of the computation. Grid points amounting to 100×201 are used for the actual calculation. The distance between neighbouring grid points is denoted by d. As before, only a quarter of the space is subject to the relaxation computation because of the symmetry.

Single cylindrical shell

Figure 2.57 shows the distribution of equipotential lines in the rz plane for a shell of which the length is 100, the inner radius 45 and the

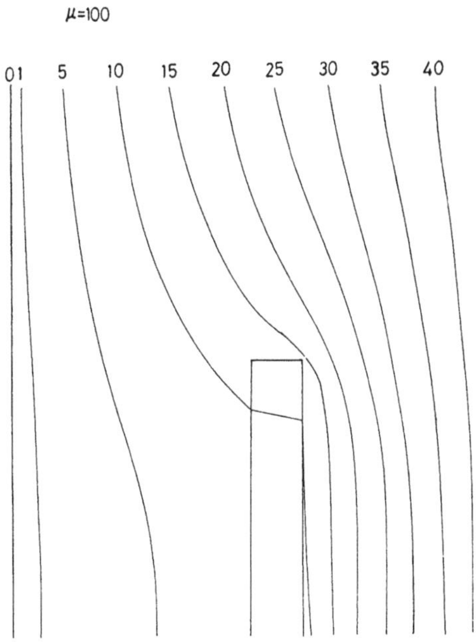

FIG. 2.57 Equipotential lines in the rz plane for a cylindrical shell of which the length is 100, inner radius 45 and outer radius 55, in units of d, and the magnetic permeability is 100.

outer radius is 55, in units of d, and for which the magnetic permeability is 100. The magnetic potential in the $z=0$ plane is calculated by means of (2.106). The equipotential lines in that plane for the above model are shown in Fig. 2.58.

Figures 2.59 and 2.60 are corresponding illustrations for a model of the same geometry but having a magnetic permeability of 1000.

As these figures show equipotential lines rather than magnetic lines of force, it is somewhat difficult to see the distribution of the magnetic fields intuitively. It is observed, however, that the density of the lines in the space around the center is smaller for the $\mu=1000$ shell than that for the $\mu=100$ shell. This suggests that the shielding effect for a shell of higher permeability is larger than that for a shell of lower permeability.

In order to see the effect of the shape of a single shell on the shielding power, the distributions of equipotential lines in the rz plane are shown for a number of models representing slim, fat, thin, thick, short and long shells, respectively. It is assumed that the permeability is always 500. The distributions are illustrated in Figs. 2.61 to 2.70.

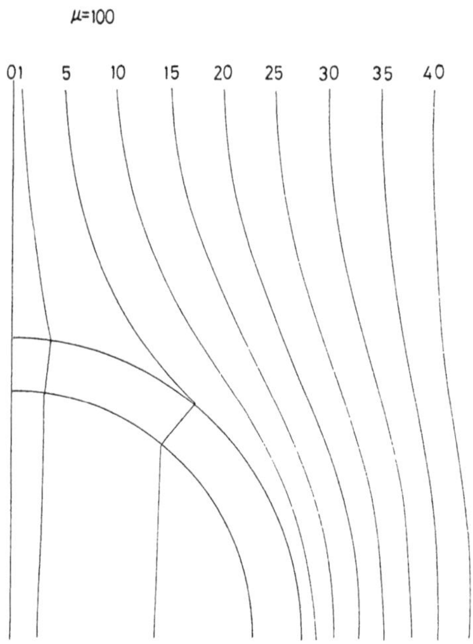

FIG. 2.58 Equipotential lines in the $z=0$ plane for a model whose dimensions are given in the caption of Fig. 2.57.

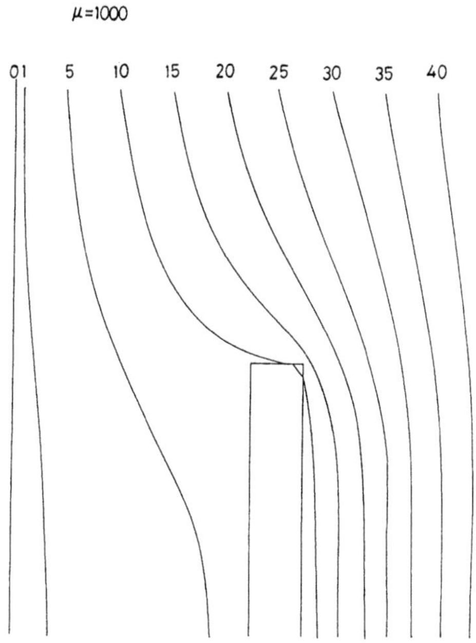

FIG. 2.59 Equipotential lines in the *rz* plane for a $\mu = 1000$ model, the geometry of which is similar to that in Fig. 2.57.

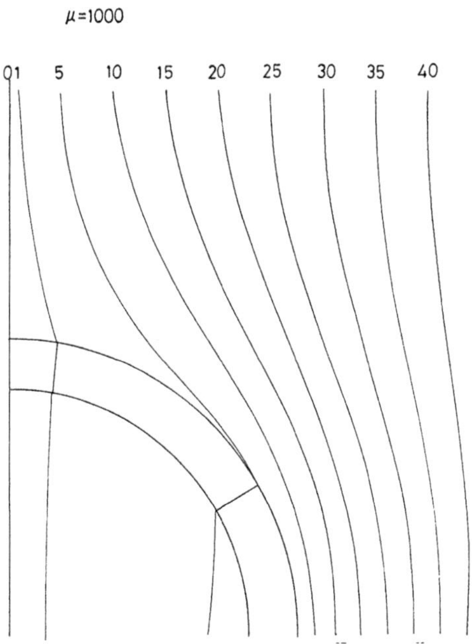

FIG. 2.60 Equipotential lines in the $z = 0$ plane for a $\mu = 1000$ model, the geometry of which is similar to that in Fig. 2.58.

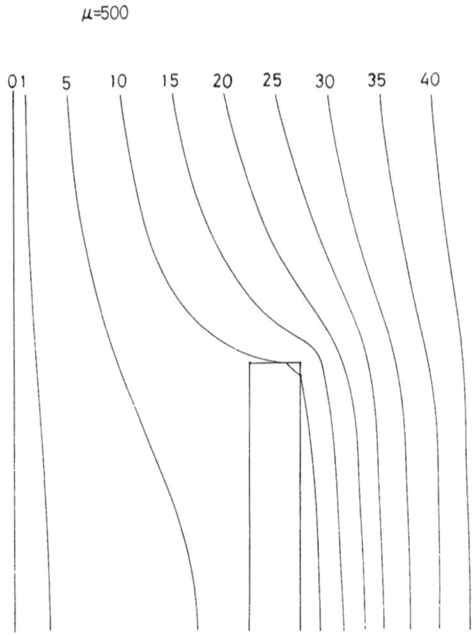

FIG. 2.61 Equipotential lines in the rz plane for a model shown in the figure. $\mu = 500$ is assumed.

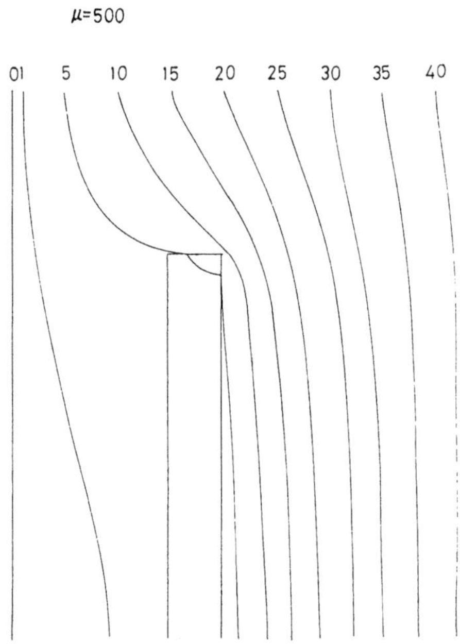

FIG. 2.62 Equipotential lines in the rz plane for a model shown in the figure. $\mu = 500$ is assumed.

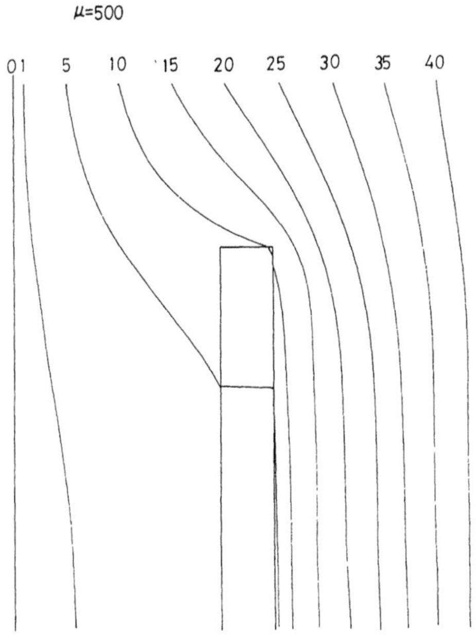

FIG. 2.63 Equipotential lines in the *rz* plane for a model shown in the figure. $\mu = 500$ is assumed.

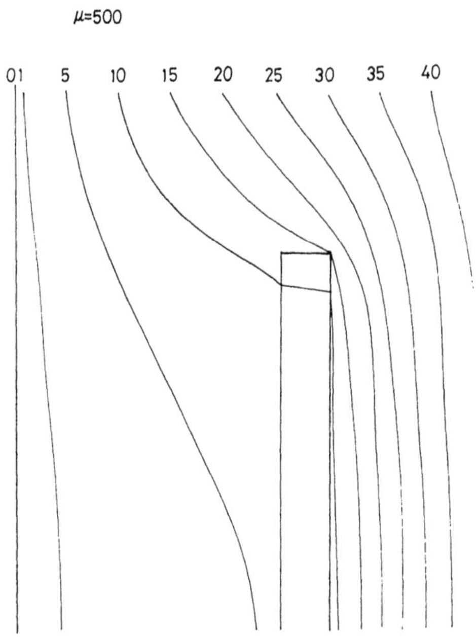

FIG. 2.64 Equipotential lines in the *rz* plane for a model shown in the figure. $\mu = 500$ is assumed.

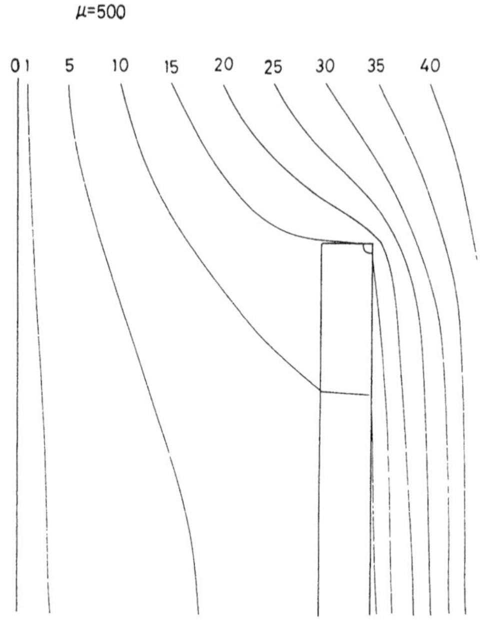

FIG. 2.65 Equipotential lines in the rz plane for a model shown in the figure. $\mu=500$ is assumed.

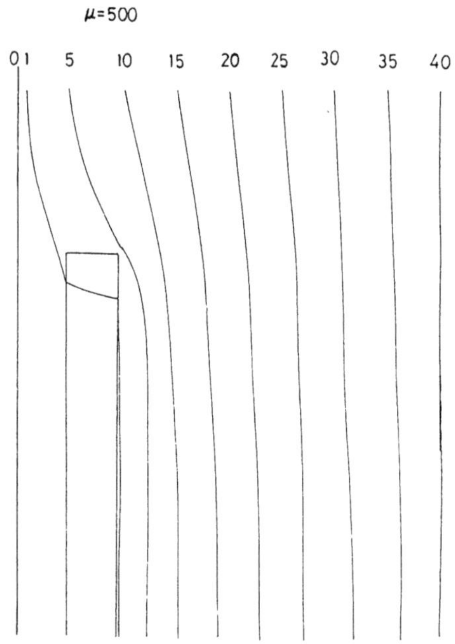

FIG. 2.66 Equipotential lines in the rz plane for a model shown in the figure. $\mu=500$ is assumed.

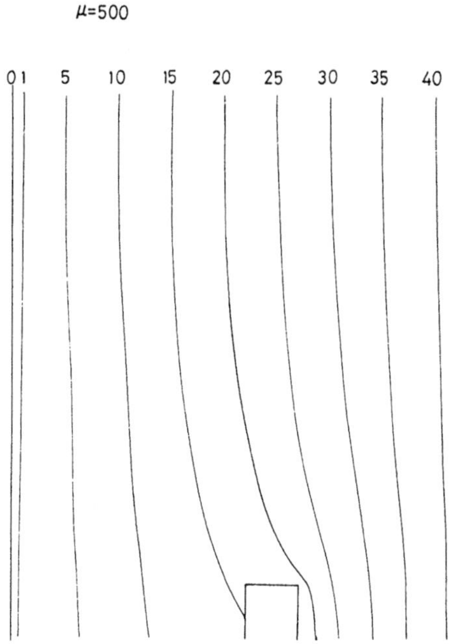

FIG. 2.67 Equipotential lines in the rz plane for a model shown in the figure. $\mu=500$ is assumed.

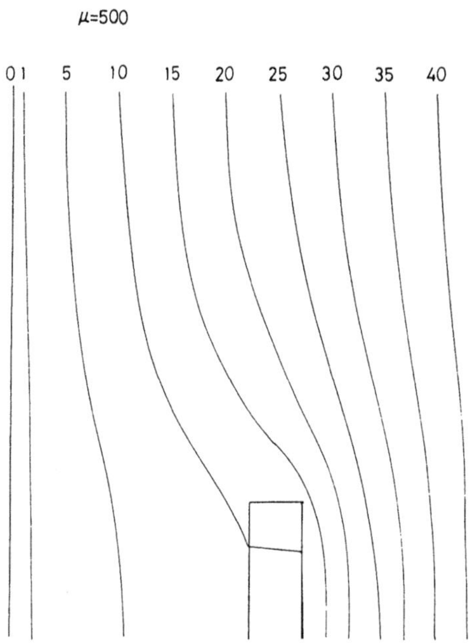

FIG. 2.68 Equipotential lines in the rz plane for a model shown in the figure. $\mu=500$ is assumed.

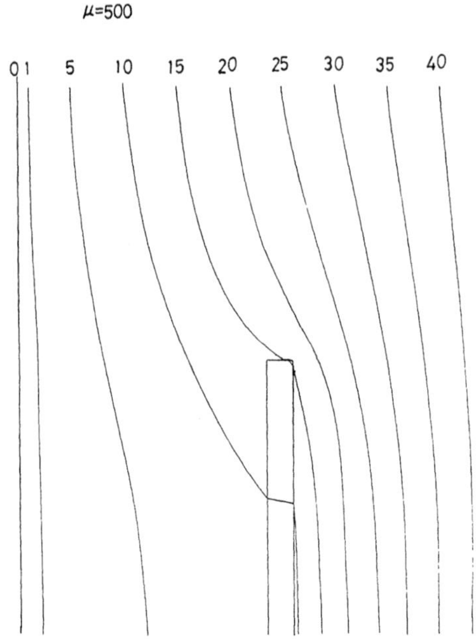

FIG. 2.69 Equipotential lines in the rz plane for a model shown in the figure. $\mu=500$ is assumed.

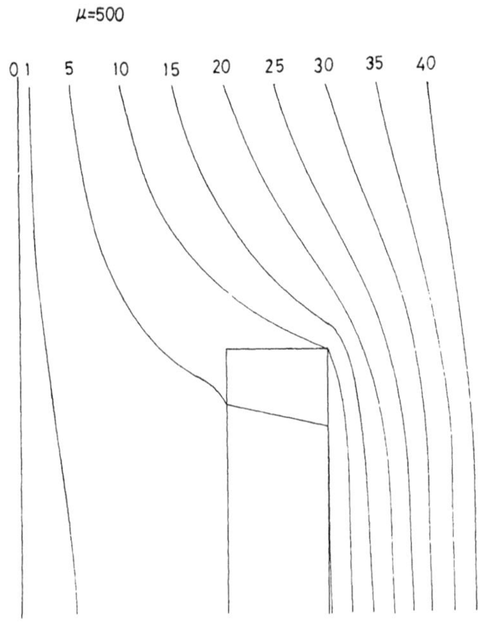

FIG. 2.70 Equipotential lines in the rz plane for a model shown in the figure. $\mu=500$ is assumed.

Looking at the equipotential lines thus obtained, it is readily observed that the longer or the thicker the shell, the more notable is the rarefaction of equipotential lines in the central space.

The components of magnetic flux density are given by

$$\begin{aligned} B_x &= -\mu \frac{\partial W}{\partial x} \\ &= -\mu \left(\cos^2\theta \frac{\partial \phi}{\partial r} + \frac{\sin^2\theta}{r} \phi \right) \\ B_y &= -\mu \frac{\partial W}{\partial y} \\ &= -\mu \left(\sin\theta \cos\theta \frac{\partial \phi}{\partial r} + \frac{\sin\theta\cos\theta}{r} \phi \right) \\ B_z &= -\mu \frac{\partial W}{\partial z} \\ &= -\mu \cos\theta \frac{\partial \phi}{\partial z} . \end{aligned} \tag{2.127}$$

The degree of shielding by single shells is here defined by the shielding ratio which is given as

$$S_i = B/B_0 \tag{2.128}$$

where

$$B = (B_x^2 + B_y^2 + B_z^2)^{1/2} \tag{2.129}$$

while B_0 is the magnetic flux density of the external field.

In Figs. 2.71, 2.72, 2.73 and 2.74, shielding ratio are shown for various single shell models. The values of S_i are shown for $r=0$, 10, 20, 30, 40 and 50 in units of d along the $\theta=0$ axis in the $z=0$ plane. In these figures, it is observed that S_i at the central axis is larger than at points nearer the shell wall. This means that the shielding effect is larger at a point nearer the shell wall than at the center. It is of course obvious from the figures of magnetic equipotential lines that the shielding effect along the z axis becomes largest at the center of the shell.

The top four plots in Fig. 2.71 show how S_i changes when the radius of the shell changes keeping the thickness and permeability constant. In the figure, it is indicated that the plots by open triangles, open squares, open circles and solid rectangles correspond to the models shown in Figs. 2.62, 2.63, 2.64 and 2.65, respectively. The plots by solid circles represent S_i's for the model shown in Fig. 2.66 which corresponds to an extremely slim shell. It is apparent in Fig. 2.71 that

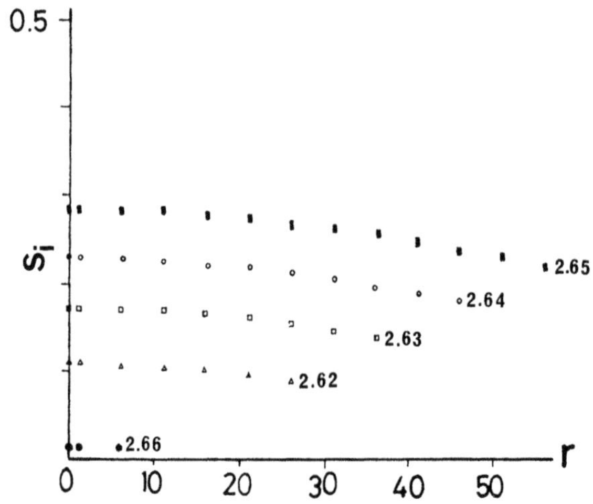

FIG. 2.71 Changes in the shielding ratio S_i with r along the $\theta=0$ axis in the $z=0$ plane for models of which the geometry and magnetic permeability are shown in the figures with corresponding numbers.

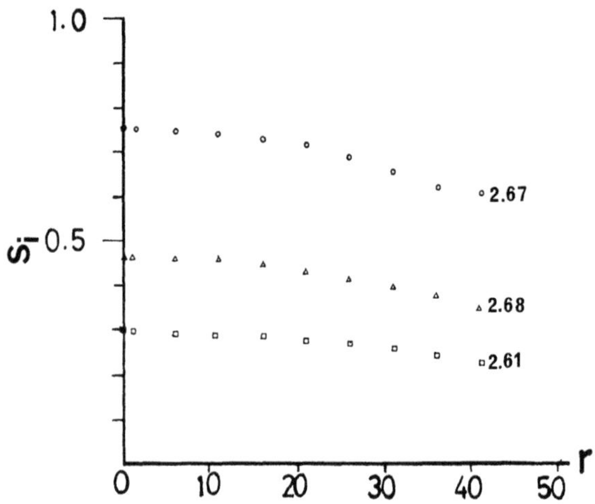

FIG. 2.72 Changes in the shielding ratio S_i with r along the $\theta=0$ axis in the $z=0$ plane for models of which the geometry and magnetic permeability are shown in the figures with corresponding numbers.

the smaller the radius of the shell, the higher is the degree of shielding. We should also note that the shielding ratios for almost all the cases are

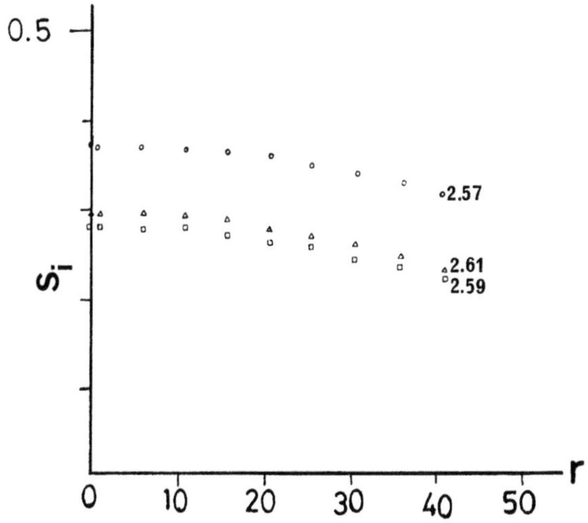

FIG. 2.73 Changes in the shielding ratio S_i with r along the $\theta=0$ axis in the $z=0$ plane for models of which the geometry and magnetic permeability are shown in the figures with corresponding numbers.

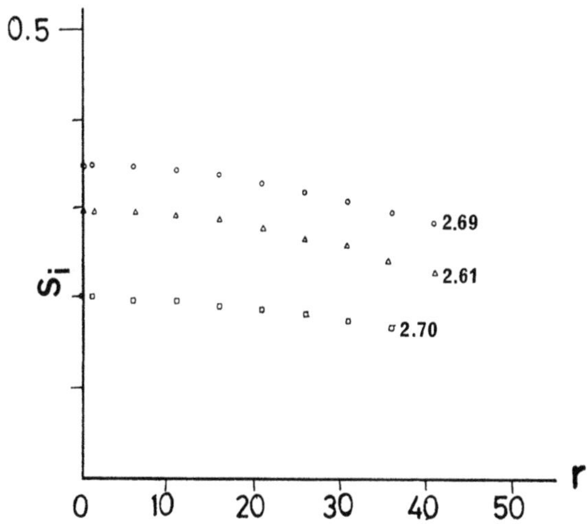

FIG. 2.74 Changes in the shielding ratio S_i with r along the $\theta=0$ axis in the $z=0$ plane for models of which the geometry and magnetic permeability are shown in the figures with corresponding numbers.

larger than for an infinitely-long shell by a factor of 10 or more. Only for the model in Fig. 2.66 is the ratio about the same as that of the infinitely-long shell. We therefore arrive at an important conclusion that a long cylindrical shell, for which the length is about ten times the radius, must be used in order to achieve a shielding power comparable to that of a shell of infinite length.

Figure 2.72 shows how the shielding ratio S_i changes when the length of the shell is changed. It is clearly demonstrated in the figure that the longer the shell, the larger the shielding effect.

Figure 2.73 is drawn specially for showing the influence of magnetic permeability on the shielding power. As expected, the shielding power is larger for higher permeabilities provided the shell geometry is the same. However, there is not much difference in the shielding power between shells of $\mu=500$ and $\mu=1000$ although the difference is considerable between shells of $\mu=100$ and $\mu=500$ or 1000.

The differences in the shielding power between shells having different thickneses are demonstrated well in Fig. 2.74. Although it is shown in the figure that the thicker the shell, the larger is the shielding effect, it should be pointed out that, in order to have a higher shielding power, it is better to use a longer cylindrical shell rather than to use a thicker shell.

Double cylindrical shells

The above study on rearrangement of magnetic equipotential lines in the presence of a single shell can readily be extended to cases of double shells. Figures 2.75 and 2.76 show the distributions of magnetic equipotential lines, respectively, in the *rz* and $z=0$ planes. Figures 2.77 and 2.78 are similar distributions for a model of double shells for which the free space between the inner and outer shells is wider than that of the former model shown in Figs. 2.75 and 2.76. $\mu=500$ is assumed for both models.

That the degree of rarefaction of magnetic equipotential lines is a little lower inside the inner shell for the model in Fig. 2.75 than for that in Fig. 2.77 can be seen in the figures.

The shielding ratios for the two models are shown in Fig. 2.79 in which we see that the shielding power of the model shown in Fig. 2.77 is in fact higher than that of the model shown in Fig. 2.75. For comparison, similar plots for a single-shell model shown in Fig. 2.61 are also included in the figure.

Cylindrical shell with a lid

Modification of magnetic equipotential lines by cylindrical shells with a complete (Fig. 2.80) or incomplete (Figs. 2.81 and 2.82) lid at the end is shown in the respective figures. The magnetic permeability is

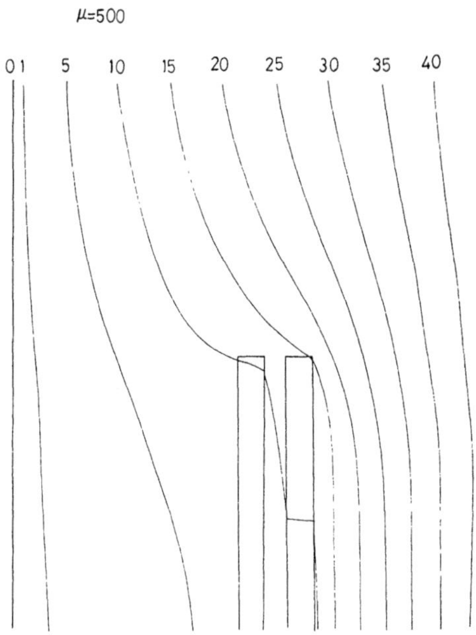

FIG. 2.75 Equipotential lines in the *rz* plane for a double cylindrical shell.

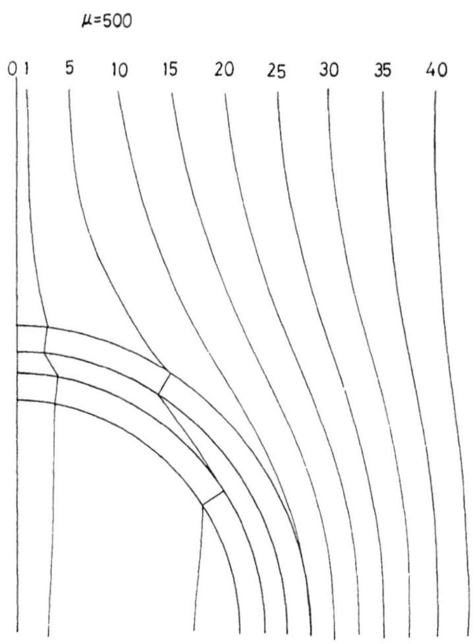

FIG. 2.76 Equipotential lines in the $z=0$ plane for a double cylindrical shell.

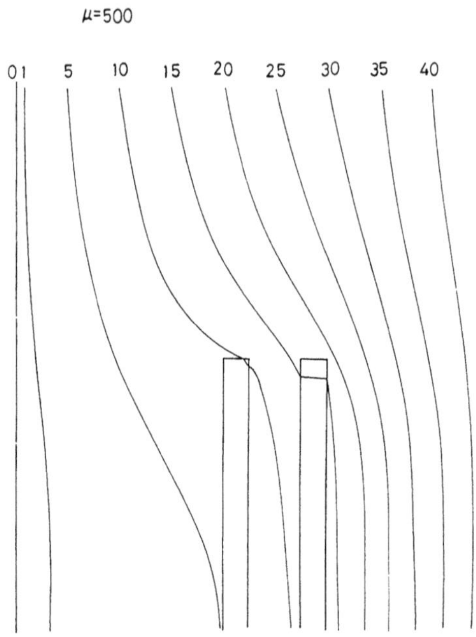

FIG. 2.77 Equipotential lines in the $rz=0$ plane for a double cylindrical shell.

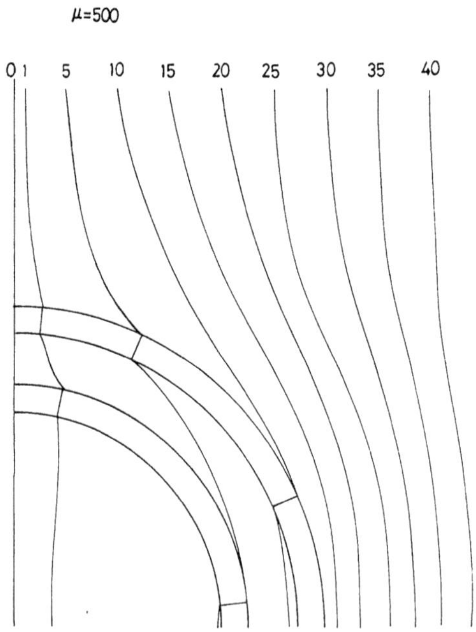

FIG. 2.78 Equipotential lines in the $z=0$ plane for a double cylindrical shell.

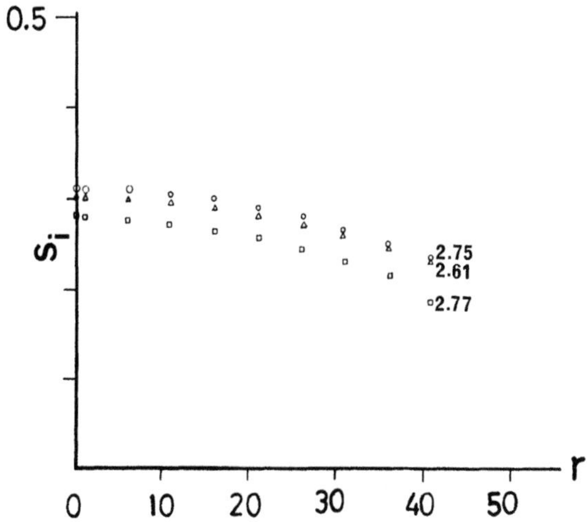

FIG. 2.79 Changes in the shielding ratio S_i with r along the $\theta=0$ axis in the $z=0$ plane for the two models shown in Figs. 2.75 and 2.77. Similar changes for the single shell model in Fig. 2.61 are also shown.

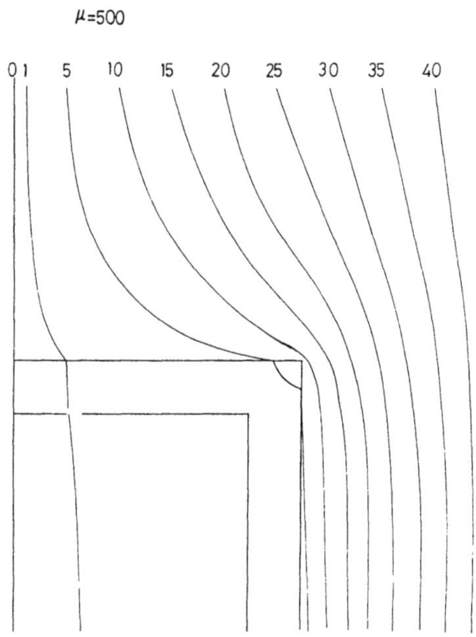

FIG. 2.80 Equipotential lines in the rz plane for a cylindrical shell with a lid at its ends.

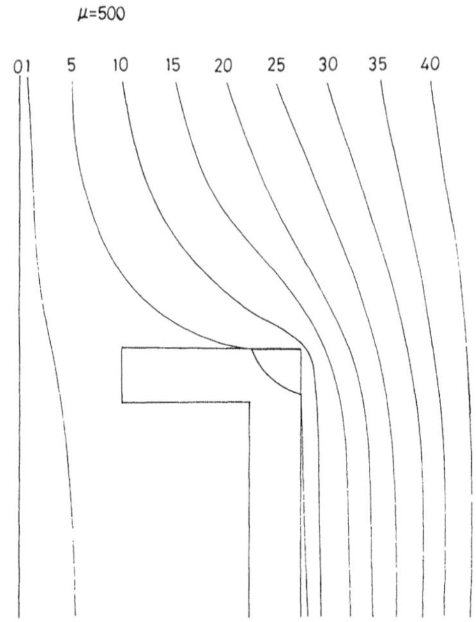

FIG. 2.81 Equipotential lines in the rz plane for a cylindrical shell with an imperfect lid at its ends.

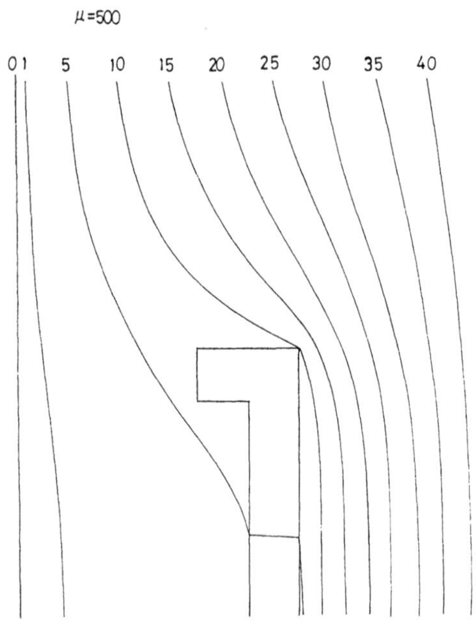

FIG. 2.82 Equipotential lines in the rz plane for a cylindrical shell with an imperfect lid at its ends.

assumed as 500 for the models.

It is remarkable that the density of magnetic equipotential lines becomes small when the lid is complete. It is also seen that the density for a model with a large hole is higher than that with a small hole. Although we have seen that a lid gives rise to a negative shielding effect in the case of an external field parallel to the cylindre axis, it turns out in the case of an external field parpendicular to the cylinder axis that a marked positive shielding effect can be provided by adding a lid to a cylindrical shell at its ends. Figure 2.83 shows S_i's for the models shown in Fig. 2.61, 2.80, 2.81 and 2.82. The shielding power of a single shell is almost doubled by adding a lid having a thickness equal to that of the shell.

Two-layer shell

In Figs. 2.84 and 2.85 are shown the magnetic equipotential lines around two-layered shells having permeabilities amounting to 100 and 1000. It is observed in the figures that the density of magnetic equipotential lines for the model in Fig. 2.84 is higher than for that in Fig. 2.85. We therefore see that a magnetic substance of higher permeabilty should occupy the inner layer in order to achieve a better shielding. Such a conclusion is contrary to that for a two-layered shell placed in an external magnetic field parallel to the cylinder axis as we have seen before.

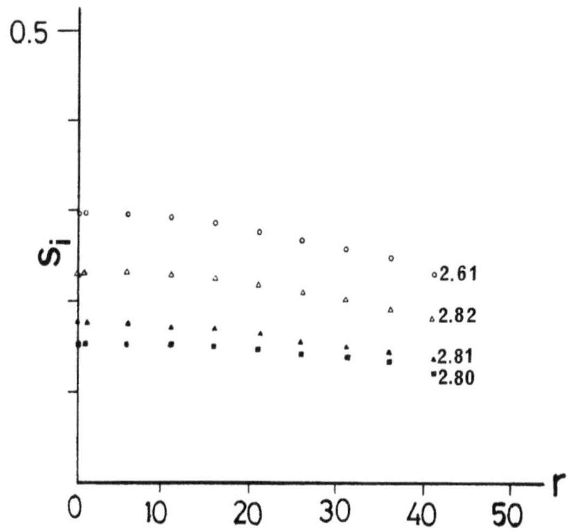

FIG. 2.83 Changes in the shielding ratio S_i with r along the θ=0 axis in Figs. 2.80, 2.81 and 2.82. Similar changes for the single shell model without a lid as in Fig. 2.61 are also shown.

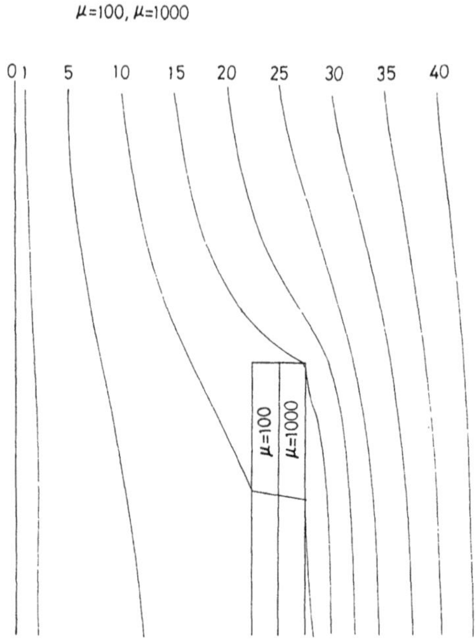

FIG. 2.84 Equipotential lines in the *rz* plane for a two-layer cylindrical shell. The magnetic permeability of the inner shell is higher than that of the outer shell.

The degree of shielding for the models is shown in terms of S_i in Fig. 2.86.

2.3.2.3 *Summary of magnetic shielding by cylindrical shells of finite length*

All through the studies conducted in this subsection, it is concluded that a very long cylindrical shell should be used in order to have a shielding power comparable to that of a shell of infinite length. The ratio of length to radius should be ten or more for achieving such a shielding power in the case of a cylindrical shell having a permeability in the range 100–1000.

When the length-to-radius ratio is around 2, the external field can be reduced to one tenth of its original intensity at the center of a cylindrical shell having a thickness amounting to one fifth or one tenth of the radius and a permeability of 100– 1000.

There are considerable differences in the mode of shielding when the external field is perpendicular and when it is parallel to the cylinder axis, although no details are given in this summary. The details are to be found in 2.3.2.1 and 2.3.2.2 of this subsection.

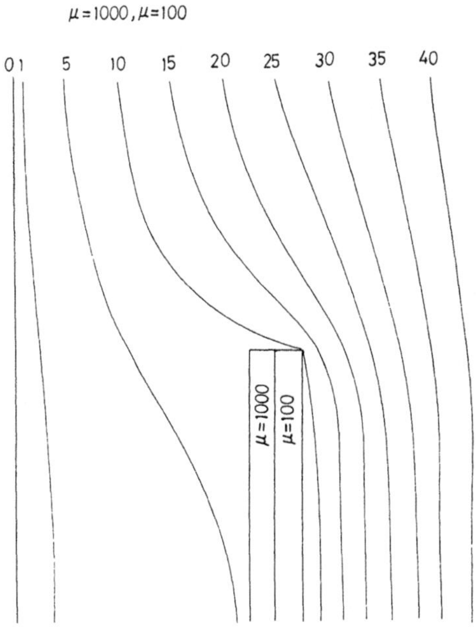

FIG. 2.85 Equipotential lines in the rz plane for a two-layer cylindrical shell. The magnetic permeability of the inner shell is higher than that of the outer shell.

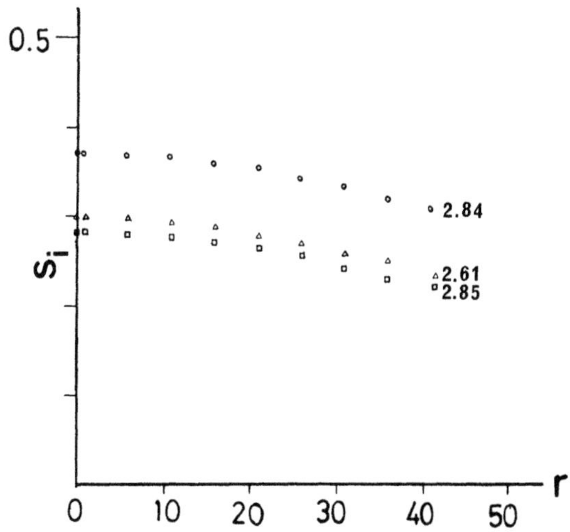

FIG. 2.86 Changes in the shielding ratio S_i with r along the $\theta=0$ axis in the $z=0$ plane for the two-layer models shown in Figs. 2.84 and 2.85. Similar changes for the single shell model of Fig. 2.61 are also shown.

2.4 *Magnetic Shielding by Rectangular Shells of Infinite Length*

The theory of magnetic shielding by spherical and cylindrical shells as studied in the foregoing subsections can readily be extended to solid and hollow prisms having a square (or rectangular) cross-section as long as the length of the prism is assumed to be infinite.

2.4.1 *Models and magnetic lines of force*

Let us take a prism having a magnetic permeability μ_2 as shown in Fig. 2.87. Although the prism is solid in the figure, it may become a hollow one as shown later. The prism is surrounded by a medium having a permeability μ_1. It is assumed that there is a uniform magnetic field parallel to the x direction at infinity.

For the purpose of applying the relaxation method to the problem, the xy plane is covered by a square net of which the grid interval is d. In practice the number of grid points amounts to 161×161 although actual calculations are performed only over a quarter or half of the space taking symmetry into account.

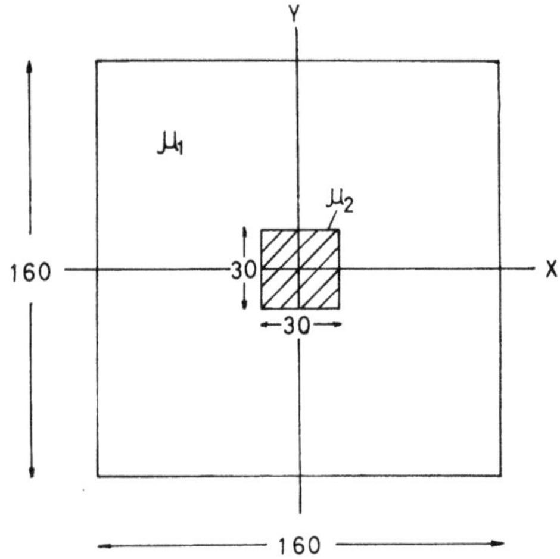

FIG. 2.87 An infinitely-long square prism of magnetic permeability μ_2 placed in a space of which the magnetic permeability is μ_1. At the outer boundary, the effect of the highly-permeable prism is assumed to vanish. Only a cross section in the xy plane is shown in the figure. The size of the prism and outer boundary are given by 30×30 and 160×160 in units of the grid interval d.

The highly-permeable matter does not occupy the space outside the central region defined by the $30d \times 30d$ area. The effect of the highly-permeable matter is assumed to vanish at $x, y=160d$.

2.4.1.1 Relaxation

The problem is here treated as two-dimensional in the xy plane as seen in Fig. 2.87. In that case we take a magnetic flux function ψ which satisfies

$$\frac{\partial^2 \psi}{\partial x^2}+\frac{\partial^2 \psi}{\partial y^2}=0 \qquad (2.130)$$

which is similar to (2.26). The components of magnetic flux density are then given by

$$B_x=-\frac{\partial \psi}{\partial y}, \quad B_y=\frac{\partial \psi}{\partial x} \qquad (2.131)$$

which is similar to (2.25).

The difference form of Eq.(2.130) becomes

$$\psi_1+\psi_2+\psi_3+\psi_4-4\psi_0=0 \qquad (2.132)$$

referring to the relaxation net shown in Fig. 2.88. Equation (2.132) is rather simpler than the equation for an axisymmetric case. The fact that the relaxation pattern is the same for all grid points makes the computation simpler.

2.4.1.2 Boundary conditions

Let us first consider a boundary shown in Fig. 2.89. ψ, ψ', μ_1 and μ_2 are the same as those defined for Fig. 2.35.

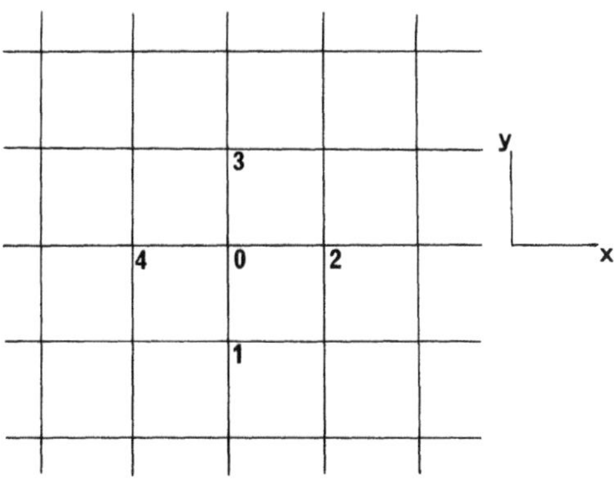

FIG. 2.88 Relaxation net in the xy plane. The distance between adjacent grid points is d.

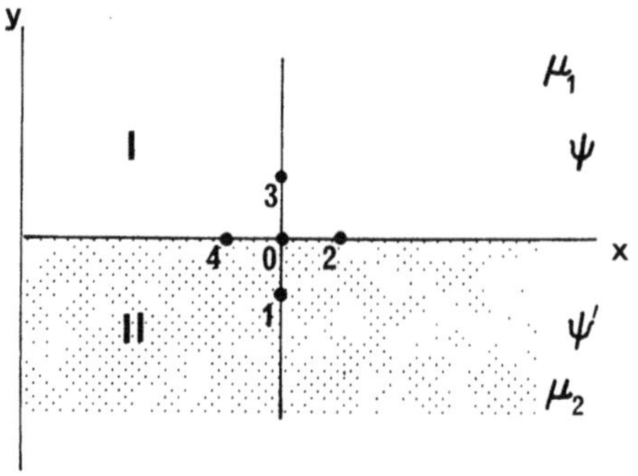

FIG. 2.89 Grid points for describing boundary conditions.

Continuity of the normal component of magnetic flux density implies that

$$\frac{\partial \psi}{\partial x}=\frac{\partial \psi'}{\partial x}$$

holds good at the boundary. On integrating both sides of the above equation with respect to x and putting the integraion constant to zero, we obtain

$$\psi=\psi'$$

at the boundary without loss of generality.

Hence, we obtain

$$\begin{aligned} \psi_2&=\psi_2' \\ \psi_0&=\psi_0' \\ \psi_4&=\psi_4' \end{aligned} \tag{2.133}$$

at the boundary.

Also, continuity of the tangential component of the magnetic field leads to

$$\frac{1}{\mu_1}\frac{\partial \psi}{\partial y}=\frac{1}{\mu_2}\frac{\partial \psi'}{\partial y}$$

at the boundary. The difference form of the equation is given as

$$\frac{1}{\mu_1}\frac{\psi_3-\bar{\psi}_1}{2d}=\frac{1}{\mu_2}\frac{\bar{\psi}_3'-\psi_1'}{2d}$$

which can be rewritten

$$T(\psi_3-\bar{\psi}_1)=\bar{\psi}_3'-\psi_1' \tag{2.134}$$

where $\bar{\psi}$ and $\bar{\psi}'$ are the hypothetical values of ψ and ψ' when media I and II are assumed to be extended beyond the boundary. T is already defined by (2.82) as

$$T=\mu_1/\mu_2\,. \tag{2.135}$$

In such hypothetical media, Eq. (2.130) should hold good, so that we have

$$\bar{\psi}_1+\psi_2+\psi_3+\psi_4-4\psi_0=0 \tag{2.136}$$

and

$$\psi_1'+\psi_2'+\bar{\psi}_3'+\psi_4'-4\psi_0'=0\,. \tag{2.137}$$

From (2.134), we obtain

$$\bar{\psi}_1=4\psi_0-\psi_2-\psi_3-\psi_4\,. \tag{2.138}$$

From (2.135) and (2.133), we obtain

$$\bar{\psi}_3'=4\psi_0-\psi_1'-\psi_2-\psi_4\,. \tag{2.139}$$

Substituting (2.138) and (2.139) into (2.134) results in

$$C\psi_1'+\psi_2+D\psi_3+\psi_4-4\psi_0=0 \tag{2.140}$$

where

$$C=\frac{2}{1+T}$$

and (2.141)

$$D=\frac{2T}{1+T}$$

as already defined in (2.89).

(2.140) provides the difference equation to be satisfied at the boundary.

When the configuration of magnetic matter is as shown in Fig. 2.90, the difference equation to be satisfied at the boundary becomes

$$D\psi_1+\psi_2+C\psi_3'+\psi_4-4\psi_0=0 \tag{2.142}$$

which can be derived in a similar manner to that for (2.140).

Next, let us consider a boundary as shown in Fig. 2.91. From the continuity of the normal component of magnetic flux density, we

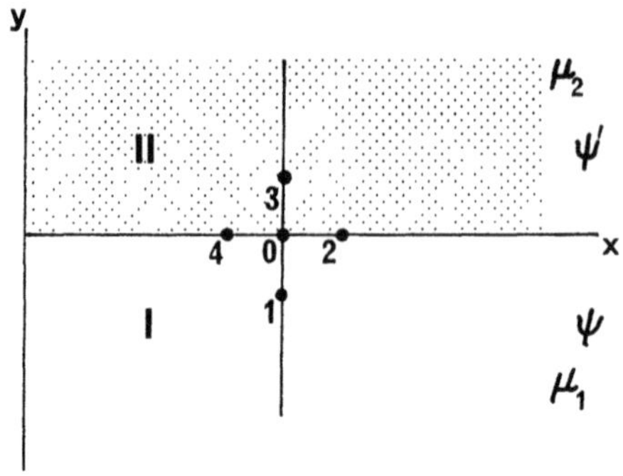

FIG. 2.90 Grid points for describing boundary conditions.

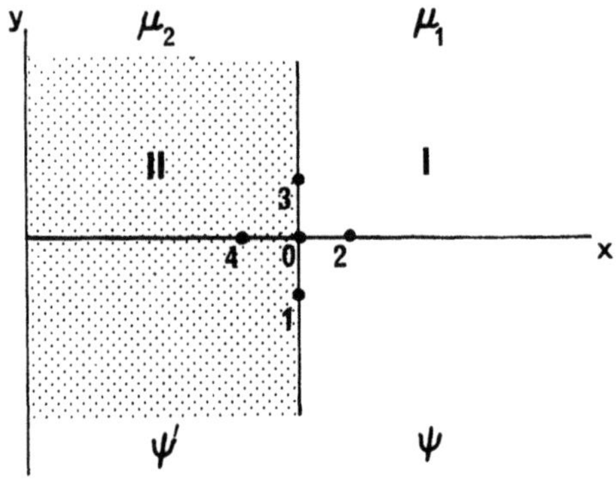

FIG. 2.91 Grid points for describing boundary conditions.

obtain

$$\frac{\partial\psi}{\partial y}=\frac{\partial\psi'}{\partial y}$$

at the boundary, so that

$$\begin{aligned}\psi_1&=\psi_1'\\ \psi_3&=\psi_3'\\ \psi_0&=\psi_0'.\end{aligned} \tag{2.143}$$

Continuity of the tangential component of the magnetic field leads to

$$\frac{1}{\mu_1}\frac{\partial\psi}{\partial x}=\frac{1}{\mu_2}\frac{\partial\psi'}{\partial x}$$

for which the difference form is

$$\frac{1}{\mu_1}\frac{\psi_2-\overline{\psi}_4}{2d}=\frac{1}{\mu_2}\frac{\overline{\psi}_2'-\psi_4'}{2d}$$

which may be further rewritten as

$$T(\psi_2-\overline{\psi}_4)=\overline{\psi}_2'-\psi_4' \qquad (2.144)$$

by taking (2.135) into account. $\overline{\psi}_4$ and $\overline{\psi}_2'$ in the above expressions are the hypothetical values in the respective regions beyond the boundary as already mentioned when we dealt with (2.134).

As we know that

$$\psi_1+\psi_2+\psi_3+\overline{\psi}_4-4\psi_0=0 \qquad (2.145)$$

and

$$\psi_1'+\overline{\psi}_2'+\psi_3'+\psi_4'-4\psi_0'=0 \qquad (2.146)$$

hold good respectively in media I and II, we obtain

$$\overline{\psi}_4=4\psi_0-\psi_1-\psi_2-\psi_3 \qquad (2.147)$$

$$\overline{\psi}_2'=4\psi_0-\psi_1-\psi_3-\psi_4' \qquad (2.148)$$

with the aid of (2.143).

Eliminating $\overline{\psi}_4$ and $\overline{\psi}_2'$ by substituting (2.147) and (2.148) into (2.144), we obtain

$$\psi_1+D\psi_2+\psi_3+C\psi_4'-4\psi_0=0 \qquad (2.149)$$

where C and D are defined in (2.141).

(2.149) is the difference equation which should be satisfied at the boundary.

When the order between media I and II is reversed as shown in Fig. 2.92, we readily see that the difference equation to be satisfied at the boundary becomes

$$\psi_1+C\psi_2'+\psi_3+D\psi_4-4\psi_0=0. \qquad (2.150)$$

In the present problem, it is assumed that a uniform magnetic field $-H_0$ parallel to the x axis exists at a great distance from the highly-permeable region. The magnetic flux function is therefore given as

$$\psi=H_0y \qquad (2.151)$$

at infinity, when $\mu_1=1$ is assumed. This provides the boundary

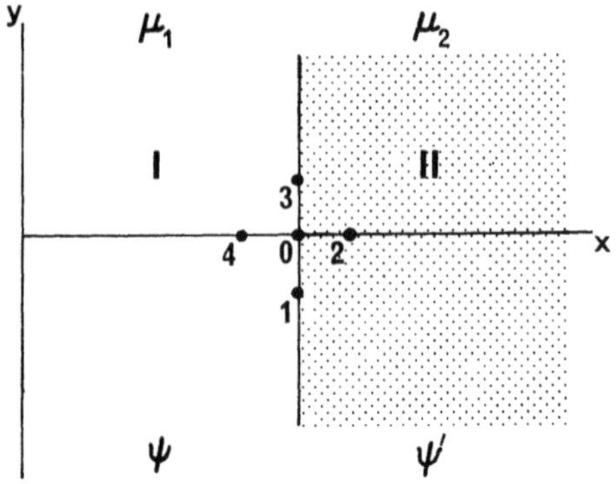

FIG. 2.92 Grid points for describing boundary conditions.

condition at a great distance where the effect of the highly-permeable matter is assumed to vanish.

2.4.2 *Solid and hollow prisms*

Based on the equations for the two media and boundaries, as developed in the last subsection, distributions of magnetic lines of force in and around solid and hollow prisms will be obtained in this subsection by means of the relaxation method.

First of all, magnetic lines of force relating to highly-permeable solid prisms, of which the permeabilities are 100, 500 and 1000, are shown in Figs. 2.93, 2.94 and 2.95, respectively. The permeability in free space is taken as unity. In these figures, we see that the external magnetic lines of force are attracted towards the prism as expected. The distribution pattern of the magnetic lines of force is much the same as that for a circular cylinder of infinite length as given in Fig. 2.5. It is also noticed that there is little difference in the distribution pattern between prisms having different permeabilities although the concentration of magnetic lines of force is more marked for the prism having higher permeability.

Figures 2.96, 2.97, 2.98 and 2.99 are examples of distribution patterns of magnetic lines of force for a number of hollow square prisms or square shells. We see clearly in these figures that the magnetic lines of force, that enter the highly-permeable shell, tend to run along the shell wall, although some leakage of the magnetic lines of force into the inner space is inevitable. The magnetic flux density therefore becomes low in the central portion of the inner space, so that some

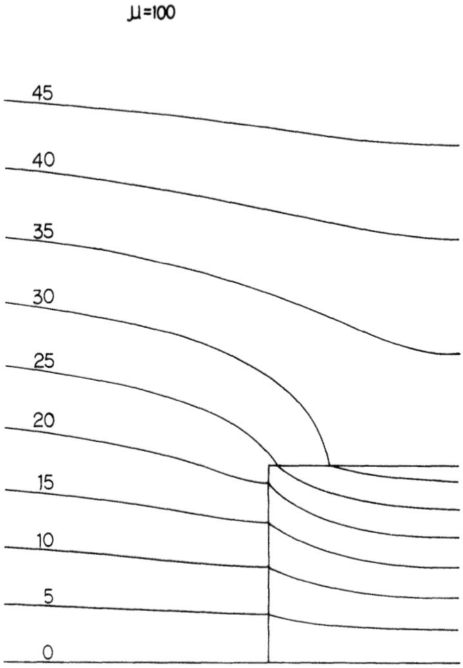

FIG. 2.93 Magnetic lines of force around a square prism of $\mu=100$.

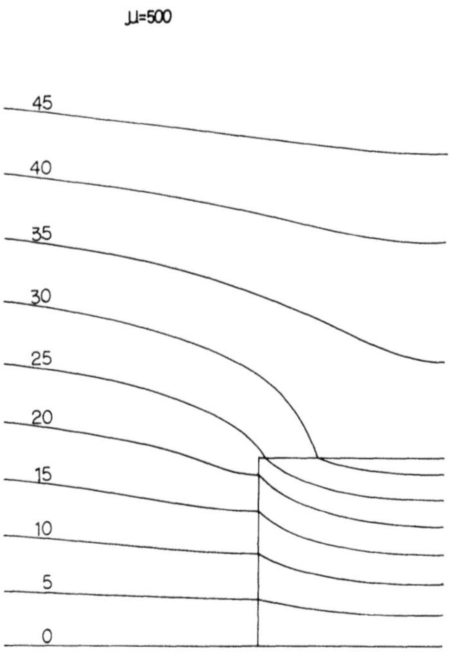

FIG. 2.94 Magnetic lines of force around a square prism of $\mu=500$.

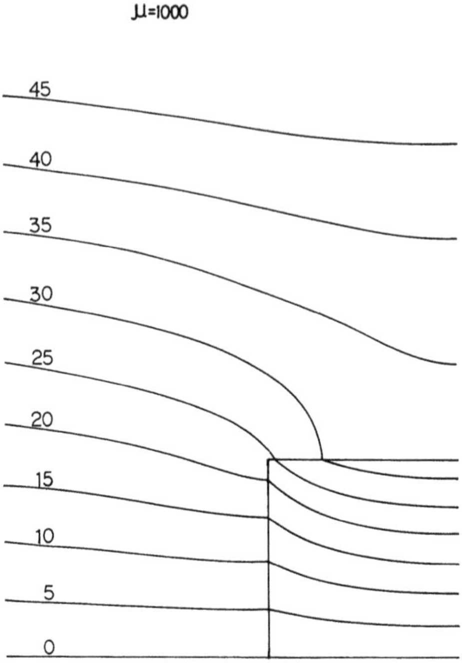

FIG. 2.95 Magnetic lines of force around a square prism of $\mu=1000$.

shielding from the external magnetic field is achieved there.

In Subsection 2.2.1, we saw that most of the magnetic lines of force run through a highly-permeable spherical shell as demonstrated, for example, in Figs. 2.12 and 2.14. In that case, the leakage of magnetic lines of force into the inner space is small, so that efficient shielding from the external field can be achieved. In contrast to those spherical shells, the leakage of magnetic line of force into the inner space for rectangular shells is considerable as we can see in Figs. 2.96, 2.97, 2.98 and 2.99, especially around the corners.

In order to see the behaviour of magnetic lines of force around a corner in more detail, the distribution of magnetic lines of force in the inner space is shown in more detail in Figs. 2.100, 2.101 and 2.102 for $\mu=100$, 500 and 1000, respectively. The thickness of the shell wall is the same for the three models. It is clearly observed in the figures that the magnetic lines of force leak out from the wall perpendicular to the external magnetic field and that they are pulled by another wall towards it, so that the magnetic lines of force in the central region are

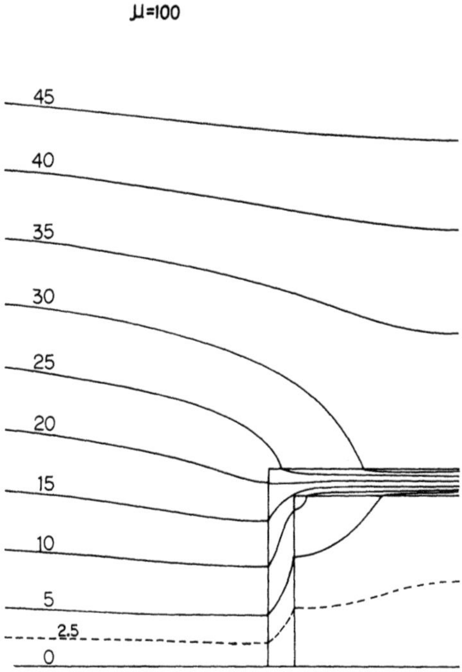

FIG. 2.96 Magnetic lines of force around a square shell of $\mu=100$.

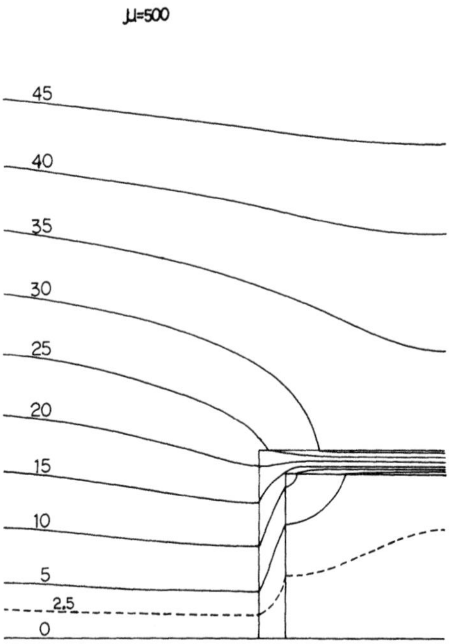

FIG. 2.97 Magnetic lines of force around a square shell of $\mu=500$.

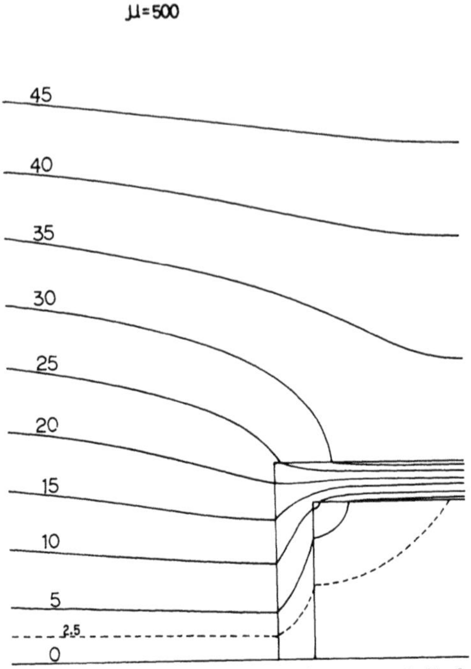

FIG. 2.98 Magnetic lines of force around a square shell of $\mu=500$.

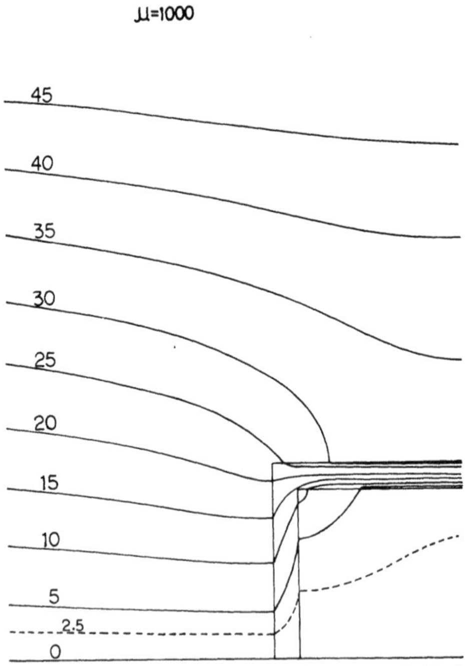

FIG. 2.99 Magnetic lines of force around a square shell of $\mu=1000$.

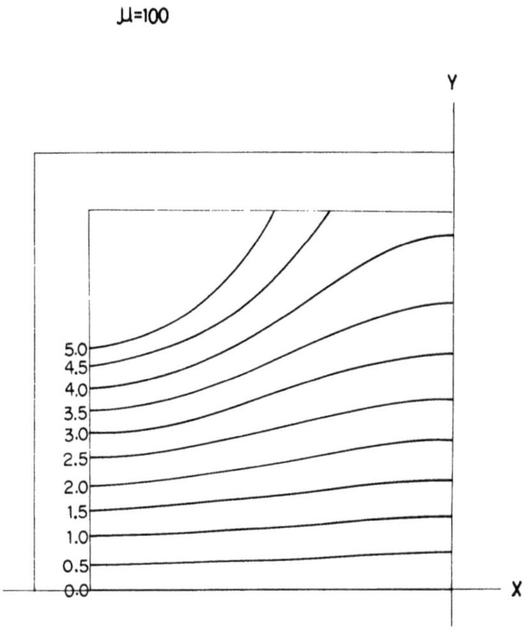

FIG. 2.100 Detailed distribution of magnetic lines of force within a square shell of $\mu=100$.

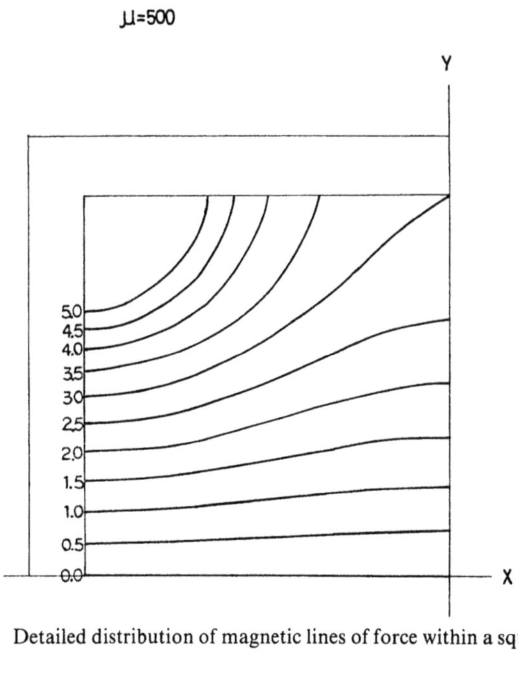

FIG. 2.101 Detailed distribution of magnetic lines of force within a square shell of $\mu=500$.

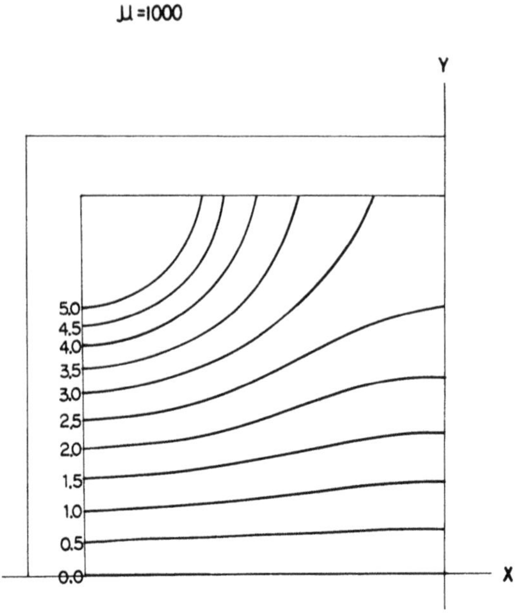

FIG. 2.102 Detailed distribution of magnetic lines of force within a square shell of $\mu=100$.

rarefied. The degree of rarefaction becomes higher as the permeability becomes higher. In Figs. 2.100, 2.101 and 2.102, we also notice that the magnetic lines of force are less dense near the upper wall than in the central region, so that the magnetic flux density near the wall is lower than that near the center.

The magnetic flux densities are then obtained on the y axis from the first equation of (2.131). The differentiation is performed numerically. It is obvious that the magnetic flux density or magnetic field on the y axis is parallel to the external magnetic field. Figure 2.103 shows the ratio of the magnetic field intensity at various points on the y axis to that of the ambient external magnetic field for $\mu=100$, 500 and 1000, respectively. The ratio is equivalent to the shielding ratio or S_i as used by the author in the previous sections.

The ratio of the inner length to the outer length of the shell for the model studied in Figs. 2.100, 2.101, 2.102 and 2.103 amounts to 13/15. Let us now calculate S_i's for a cylindrical shell of infinite length assuming $r_2/r_1=13/15$ in (2.72). Corresponding to $\mu=100$, 500 and 1000, S_i is obtained, respectively, as 0.141, 0.0313 and 0.0158.

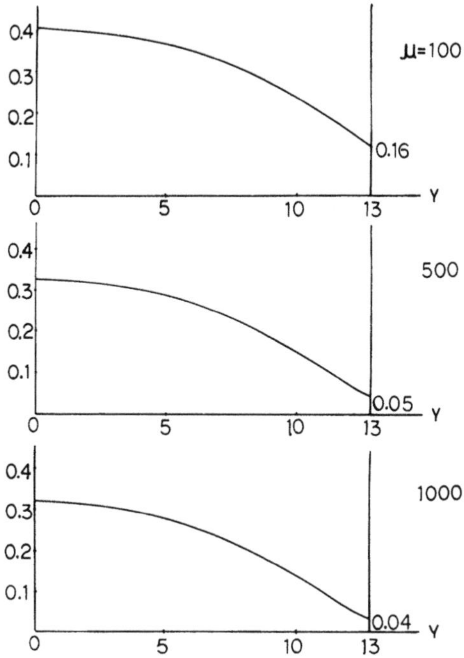

FIG. 2.103 Changes in the shielding ratio S_i along the y axis for the square shell models shown in Figs. 2.100, 2.101 and 2.102, corresponding to $\mu=100$, 500 and 1000. The distance is measured in units of d.

Meanwhile, S_i's for the square shell models become 0.41, 0.33 and 0.32, respectively, for $\mu=100$, 500 and 1000 at the center of the inner space. We therefore see that the shielding ratios for a square shell are considerably larger than those for a cylindrical shell. Such a tendency becomes more marked when the permeability is higher.

Judging from the fact that the leakage of magnetic lines of force into the inner space is considerably greater for the square shell than for the cylindrical shell, it is quite understandable that the shielding is less efficient for the former than the latter. It is concluded, therefore, that we have to use a shell of highly-permeable magnetic matter having a smooth shape in order to attain high magnetic shielding. Shells having pointed corners should not be used because magnetic lines of force leak out from there resulting in inefficient shielding.

Returning to Fig. 2.98, in which the thickness of the shell wall is 1.5 times that of the model shown in Fig. 2.97, we see that the rarefaction of magnetic lines of force is more marked for the model having a thicker wall. Hence, the shielding effect becomes larger for a shell having a thicker wall than that having a thinner wall for the same permeability.

2.4.3 *Incomplete square shell*

The relaxation analyses for obtaining distributions of magnetic lines of force in and around square shells can readily be extended to cases in which some portions of the shell are missing. As typical examples, the two models b and c, shown in the middle and bottom of Fig. 2.104, will be studied in this subsection. Model a in the figure is a square shell having no hole in the wall, and was studied in detail in the last subsection.

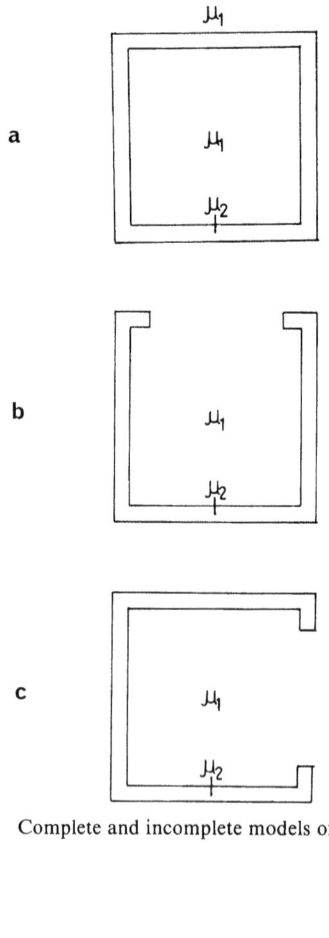

FIG. 2.104 Complete and incomplete models of square shell.

The distributions of magnetic lines of force around models b and c are illustrated in Figs. 2.105 and 2.106, respectively. Although only the models for which the permeability is 500 are here considered, there is no difficulty in performing similar relaxation analyses for models having different permeabilities.

Looking at these figures, we see that considerable concentration of the magnetic lines of force takes place in the highly-permeable wall as in the case of a complete shell studied in the last subsection. Accordingly, rarefaction of magnetic lines of force is observed in the inner space.

In the case of model b, in which some part of the wall parallel to the external magnetic field is missing, we see, however, some penetration of magnetic lines of force into the inner space through the hole. It is therefore certain that the shielding effect for model b will not be as good as for model a, the complete square shell. The magnetic fields on the y axis are estimated by numerically differentiating the magnetic flux function as before. The top half of Fig. 2.107 shows the magnetic fields

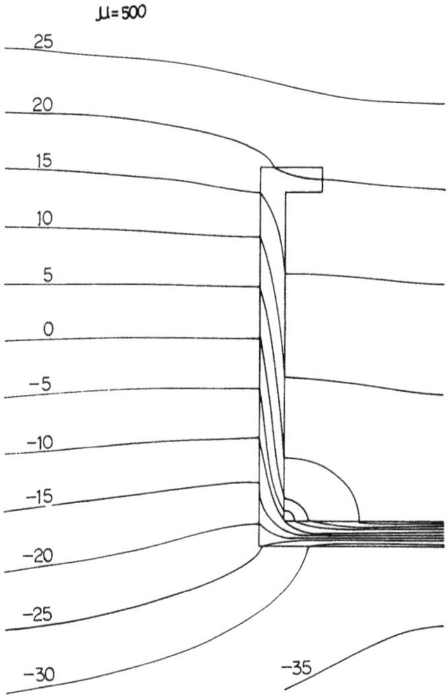

FIG. 2.105 Magnetic lines of force around the incomplete square shell model b shown in Fig. 2.104. $\mu=500$ is assumed.

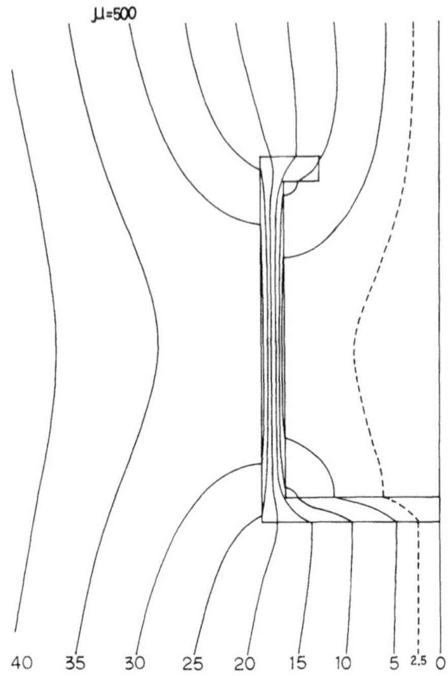

FIG. 2.106 Magnetic lines of force around the incomplete square shell model c shown in Fig. 2.104. $\mu=500$ is assumed.

for positive values of y on the y axis. The field is measured in units of the external field, so that the ordinate is equivalent to S_i. Meanwhile, the solid curve in the bottom graph shows how S_i changes with negative values of y on the y axis. The corresponding curve for model a, the complete shell, is shown as a dashed curve in the figure, too. We therefore see that the degree of shielding of model b is lower by a factor of about 30% than that of model a in the central space.

The distribution of S_i on the y axis for model c is similarly estimated and shown in Fig. 2.108 in which we see that there is very little difference in the shielding power in the central space between models c and a, the incomplete and complete shells.

The above conclusions emphasize the importance of walls parallel to the external magnetic field for achieving a shielding power of high degree. On the contrary, walls perpendicular to the external field play a relatively unimportant role as far as shielding is concerned. Unlike the cases for cylindrical or spherical shells, the magnetic lines of force

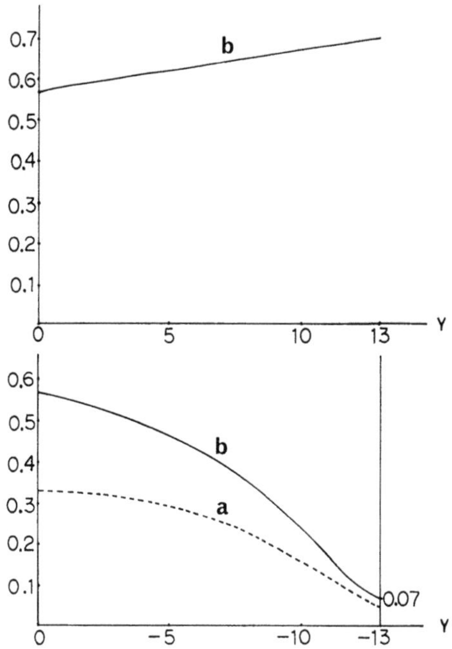

FIG. 2.107 Changes in S_i along the y axis for the model b shown in Fig. 2.104. The distance is measured in units of d. Similar changes in S_i for the model a are also shown.

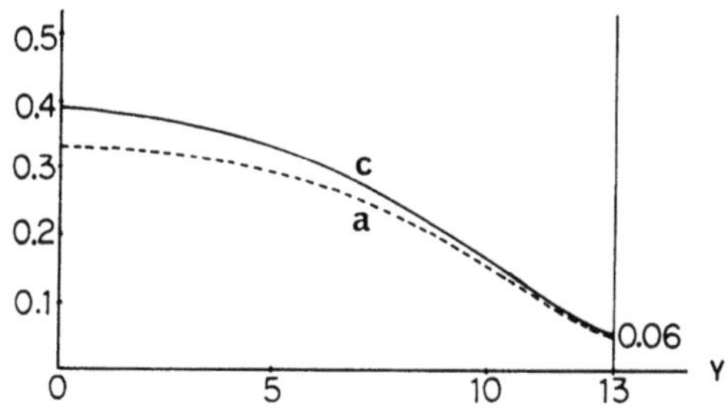

FIG. 2.108 Changes in S_i along the y axis for the model c shown in Fig. 2.104. The distance is measured in units of d. Similar changes in S_i for the model a are also shown.

which enter the highly-permeable wall cannot run along the curvature of the shell wall and therefore leak out into the inner space. As the wall perpendicular to the external field tends to attract the magnetic lines of force of the external field towards it, it even plays a negative role respect of the shielding.

In light of the above discussion, it may be concluded that the best way to accomplish shielding by a rectangular shell of infinite length is to make the length of the shell wall parallel to the external magnetic field as long as possible. The shell walls perpendicular to the external field would be better removed. Such a conclusion is much the same as that in the case of cylindrical shells of finite length when the external field is parallel to the cylinder axis.

It should be borne in mind, however, that the conclusion would be different for a rectangular or square shell when the length of the shell is finite, although no quantitative study of such a case has been carried out because the relaxation analysis for a three-dimentional case is so complicated and difficult that it is beyond the capacity of the computer available for the present study.

Chapter 3

ELECTROMAGNETIC SHIELDING

When an electric conductor, which may or may not be magnetically permeable, is placed in a time-dependent magnetic field, electric currents are necessarily induced in the conductor by electromagnetic induction. The magnetic field produced by the induced electric currents tends in general to oppose the inducing magnetic field, so that the magnetic field inside a conductor, or a space surrounded by a conductor, usually becomes smaller in amplitude than that of the external inducing magnetic field. This is a kind of shielding or screening which may be called electromagnetic shielding.

As it is known that the earth is surrounded by an electrically-conducting ionosphere and that the interior of the earth deeper than a certain level is also conducting, electromagnetic induction by changes in the magnetic field arising from extraterrestrial origins plays a vitally important role in studying the geomagnetic variation field. Electromagnetic induction in oceans is also important for geomagnetic variations of short period because sea water is highly conducting. It is also likely that the geomagnetic secular variation arising from some magneto-hydrodynamic process in the earth's core, which occupies that part of the earth deeper than 2900 km, is subject to some electromagnetic shielding as the variation penetrates through the earth's mantle and reaches the earth's surface.

Electromagnetic shielding is also important for eliminating electric and magnetic noise from a space in which an electronic device is installed. For example, it is often required to eliminate hum or noise arising from commercial power lines from a high-sensitivity amplifier.

It is the aim of this chapter to advance a general theory of electromagnetic shielding and to apply the theory to a number of typical problems. In addition to electromagnetic shielding by plane, spherical and cylindrical shells, which can be solved analytically, electromagnetic shielding by shells having square and octagonal cross sections will be studied by numerical methods. When the wavelength of an inducing field is much larger than the thickness of a conducting shell or sheet, the problem reduces to electromagnetic induction in an infinitesimally thin conducting sheet. A particularly successful approach can be made in the case of a thin sheet as will be seen in this chapter. The mathematics for such a case becomes far simpler than for a sheet of finite thickness.

Discussion in this chapter will be limited to electromagnetic induction by a slowly changing field. In other words the electric displacement current is ignored, so that the problem is always quasi-stationary. Accordingly, problems related to the shielding of high-frequency radio waves will not be included in this chapter.

3.1 Fundamental Equations

Let us start from Maxwell's equations, which can be written in electromagnetic units as

$$\text{curl}\,\boldsymbol{H}=4\pi\boldsymbol{i}+\partial\boldsymbol{D}/\partial t \tag{3.1}$$

and

$$\text{curl}\,\boldsymbol{E}=-\partial\boldsymbol{B}/\partial t \tag{3.2}$$

together with

$$\text{div}\,\boldsymbol{B}=0 \quad\text{and}\quad \text{div}\,\boldsymbol{D}=4\pi\rho \tag{3.3}$$

where $\boldsymbol{H}$ is magnetic field, $\boldsymbol{i}$ electric current density, $\boldsymbol{D}$ electric displacement, $\boldsymbol{E}$ electric field, $\boldsymbol{B}$ magnetic induction or magnetic flux density, and ρ volume density of electric charge. Assuming an isotropic medium, we have

$$\boldsymbol{D}=\epsilon\boldsymbol{E},\quad \boldsymbol{B}=\mu\boldsymbol{H},\quad \boldsymbol{i}=\sigma\boldsymbol{E} \tag{3.4}$$

where ϵ is dielectric constant, μ magnetic permeability and σ electric conductivity. If we take the divergence of (3.1), we have

$$4\pi\,\text{div}\,\boldsymbol{i}=-\text{div}\,\partial\boldsymbol{D}/\partial t=-4\pi\partial\rho/\partial t.$$

Hence it is obvious that the relation

$$\text{div}\,\boldsymbol{D}=-(\epsilon/\sigma)\,\partial\rho/\partial t+\boldsymbol{i}\cdot\text{grad}\,(\epsilon/\sigma)$$

holds good. The last relation implies

$$\partial\rho/\partial t+4\pi\sigma\rho/\epsilon=(\sigma/\epsilon)\,\boldsymbol{i}\cdot\text{grad}\,(\epsilon/\sigma). \tag{3.5}$$

It is therefore seen that ρ is independent of field vectors either when ϵ/σ is constant or when $\boldsymbol{i}$ is perpendicular to grad (ϵ/σ). In such cases, the initial charge is subject to an exponential decay. These are the only cases in which the field vectors are treated separately from the charge distribution. For instance, when the electric current flows along a surface in which the conductivity is constant (ϵ=constant is also assumed), such a treatment becomes possible. In general, however, it should be borne in mind that the separation of field vectors and charge distributions cannot be made.

Let us take a case in which ϵ and μ are constant. We also assume either that σ is constant or that electric currents are flowing along a surface in which the conductivity is constant. In view of the discussion in the last paragraph, we may assume $\rho=0$ and hence

$$\text{div}\,\boldsymbol{E}=0 \tag{3.6}$$

without loss of generality. Taking curl of (3.1), we obtain

$$\nabla^2\boldsymbol{H}=\mu\,(4\pi\sigma\partial/\partial t+\epsilon\partial^2/\partial t^2)\,\boldsymbol{H} \tag{3.7}$$

because of (3.6) and (grad $\sigma\cdot\boldsymbol{E}$)=0. It may also be proved that an equation similar to (3.7) holds good for $\boldsymbol{E}$.

When $\boldsymbol{H}$ does not change too rapidly, we have $\epsilon\partial\boldsymbol{H}/\partial t\ll 4\pi\sigma\boldsymbol{H}$ in a conductor, so that (3.7) reduces to

$$\nabla^2\boldsymbol{H}=4\pi\sigma\mu\partial\boldsymbol{H}/\partial t. \tag{3.8}$$

In contrast, the conductivity is so small in a dielectric that the first term in the righthand-side of (3.7) can be ignored. We then obtain

$$\nabla^2\boldsymbol{H}=\mu\epsilon\partial^2\boldsymbol{H}/\partial t^2 \tag{3.9}$$

which, being a wave equation, indicates that electromagnetic effects are propagated with a velocity equal to $(\mu\epsilon)^{-1/2}$. When we are concerned with relatively slow field variations, the time of propagation across any portion of the dielectric can be regarded as negligible compared with the time taken by the field changes at a point to become effective. We therefore ignore the righthand-side of (3.9) and obtain

$$\nabla^2\boldsymbol{H}=0\,. \tag{3.10}$$

It is obvious that $\boldsymbol{H}$ may be derived from a scalar potential W in the dielectric. We then put

$$\boldsymbol{H} = -\operatorname{grad} W \tag{3.11}$$

as already obtained in (2.4).

In the conductor, we may define a vector potential $\boldsymbol{A}$ by

$$\operatorname{curl} \boldsymbol{A} = \boldsymbol{B}. \tag{3.12}$$

In that case, (3.2) leads to

$$\boldsymbol{E} = -\partial \boldsymbol{A}/\partial t - \operatorname{grad} \varphi \tag{3.13}$$

where φ is an arbitrary function.

It can be shown that, if

$$\operatorname{div} \boldsymbol{A} + 4\pi\sigma\mu\varphi = 0, \tag{3.14}$$

the fundamental equation (3.8) can be satisfied. It is, therefore, possible to take $\varphi = 0$, which is consistent with (3.9), (3.12) and (3.14) because we see that

$$\operatorname{div} \boldsymbol{A} = 0 \tag{3.15}$$

from (3.6). (3.13) now reduces to

$$\boldsymbol{E} = -\partial \boldsymbol{A}/\partial t, \tag{3.16}$$

while $\boldsymbol{A}$ satisfies

$$\nabla^2 \boldsymbol{A} = 4\pi\sigma\mu \partial \boldsymbol{A}/\partial t. \tag{3.17}$$

Although it has been shown in this section that the field vectors can be determined independently from the space-charge distribution when σ is distributed in a special way, this does not imply that electrostatic effects are ignored. In order that the current flows parallel to the surface, a surface-charge distribution appears.

The derivation of the above equations almost entirely depends on the theory of electromagnetic induction as summarized in RIKITAKE (1966).

3.2 Plane Slab

RUNCORN (1955), HIDE and ROBERTS (1961) and RIKITAKE (1966) studied electromagnetic induction in a conducting plane slab or sheet in relation to the shielding of the geomagnetic secular variation by the earth's mantle. The following is mostly due to RIKITAKE (1966).

Let us take a slab of infinite extent as shown in Fig. 3.1. The electric conductivity and the magnetic permeability are denoted by σ and μ, respectively. From the general theory as developed in the last section, it is obvious that we can define a magnetic potential W in the non-conducting regions and that W satisfies

$$\nabla^2 W=0, \tag{3.18}$$

which is the same as (2.6).

A typical term of the solution of (3.18) is given as

$$W_{mn}=(e_{mn}e^{-kz}+i_{mn}e^{kz})V_{mn} \quad (z<-L) \tag{3.19}$$

for the upper non-conducting region. In (3.19), k is given by

$$k^2=(m^2+n^2)(2\pi/l)^2 \tag{3.20}$$

and V_{mn} is a harmonic function that satisfies

$$\frac{\partial^2 V_{mn}}{\partial x^2}+\frac{\partial^2 V_{mn}}{\partial y^2}+k^2 V_{mn}=0. \tag{3.21}$$

l is the wavelength of the electric and magnetic fields concerned. Physically, coefficients e_{mn} and i_{mn} correspond to magnetic potentials arising from origins located in the regions defined by $z<-L$ and $z>-L$, respectively.

In the lower non-conducting region, we obtain

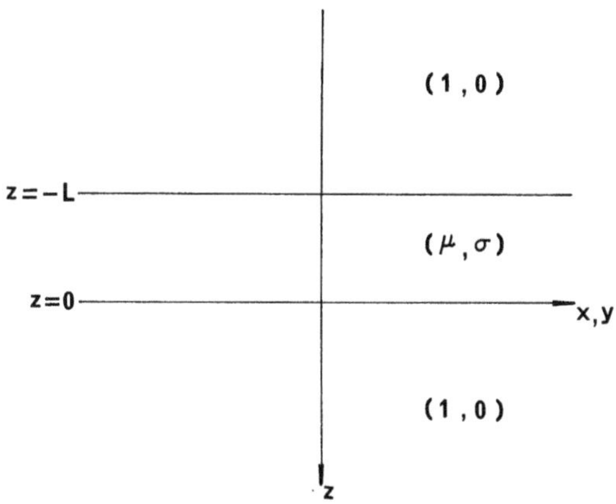

FIG. 3.1 Plane slab of infinite extent that occupies the space defined by $-L<z<0$.

$$\overline{W}_{mn}=\bar{e}_{mn}\mathrm{e}^{-kz}V_{mn} \quad (z>0) \tag{3.22}$$

on the condition that no source producing a magnetic field exists in the region.

The x, y and z components of the magnetic field are immediately obtained from (3.11) or (2.4) as

$$H=\begin{cases} -(e_{mn}\mathrm{e}^{-kz}+i_{mn}\mathrm{e}^{kz})\dfrac{\partial V_{mn}}{\partial x} \\ -(e_{mn}\mathrm{e}^{-kz}+i_{mn}\mathrm{e}^{kz})\dfrac{\partial V_{mn}}{\partial y} \\ -k(-e_{mn}\mathrm{e}^{-kz}+i_{mn}\mathrm{e}^{kz})\,V_{mn} \end{cases} \quad (z<-L) \tag{3.23}$$

$$H=\begin{cases} -\bar{e}_{mn}\mathrm{e}^{-kz}\dfrac{\partial V_{mn}}{\partial x} \\ -\bar{e}_{mn}\mathrm{e}^{-kz}\dfrac{\partial V_{mn}}{\partial y} \\ k\,\bar{e}_{mn}\mathrm{e}^{-kz}\,V_{mn} \end{cases} \quad (z>0). \tag{3.24}$$

In the conducting region, we define a vector potential $\boldsymbol{A}$ which satisfies (3.17). Replacing $\partial/\partial t$ by p, (3.17) becomes

$$(\nabla^2-\chi^2)\,\boldsymbol{A}=0 \tag{3.25}$$

where

$$\chi^2=4\pi\mu\sigma p. \tag{3.26}$$

A typical solution of (3.25) is given as

$$\boldsymbol{A}_{mn}=(C_{mn}\mathrm{e}^{-\sqrt{k^2+\chi^2}z}+D_{mn}\mathrm{e}^{\sqrt{k^2+\chi^2}z})(\boldsymbol{r}\wedge\mathrm{grad}\ V_{mn})(-L<z<0) \tag{3.27}$$

where $\boldsymbol{r}$ is the position vector connecting the origin of the coordinates to the point specified by (x, y, z). It is obvious that

$$(\boldsymbol{r}\wedge\mathrm{grad}\ V_{mn})=\begin{cases} \dfrac{\partial V_{mn}}{\partial y} \\ -\dfrac{\partial V_{mn}}{\partial x} \\ 0 \end{cases}. \tag{3.28}$$

We then obtain from (3.12) the magnetic field components as

$$H=-\mu^{-1}\begin{cases} (-C_{mn}\mathrm{e}^{-\sqrt{k^2+\chi^2}z}+D_{mn}\mathrm{e}^{\sqrt{k^2+\chi^2}z})\sqrt{k^2+\chi^2}\dfrac{\partial V_{mn}}{\partial x} \\ (-C_{mn}\mathrm{e}^{-\sqrt{k^2+\chi^2}z}+D_{mn}\mathrm{e}^{\sqrt{k^2+\chi^2}z})\sqrt{k^2+\chi^2}\dfrac{\partial V_{mn}}{\partial y} \\ (C_{mn}\mathrm{e}^{-\sqrt{k^2+\chi^2}z}+D_{mn}\mathrm{e}^{\sqrt{k^2+\chi^2}z})\,k^2V_{mn} \end{cases} (-L<z<0). \tag{3.29}$$

The continuity of the tangential component of magnetic field and the normal component of magnetic flux density at $z=-L$ and $z=0$ lead to the following simultaneous equations:

$$\begin{aligned}
&\mu\,(e_{mn}\mathrm{e}^{kL}+i_{mn}\mathrm{e}^{-kL})=\sqrt{k^2+\chi^2}\,(C_{mn}\mathrm{e}^{\sqrt{k^2+\chi^2}L}-D_{mn}\mathrm{e}^{-\sqrt{k^2+\chi^2}L})\\
&e_{mn}\mathrm{e}^{kL}-i_{mn}\mathrm{e}^{-kL}=k\,(C_{mn}\mathrm{e}^{\sqrt{k^2+\chi^2}L}+D_{mn}\mathrm{e}^{-\sqrt{k^2+\chi^2}L})\\
&\sqrt{k^2+\chi^2}\,(C_{mn}-D_{mn})=\mu\bar{e}_{mn}\\
&k\,(C_{mn}+D_{mn})=\bar{e}_{mn}\,.
\end{aligned}\qquad(3.30)$$

Solving (3.30), we obtain

$$\bar{e}/e=4\mu k\sqrt{k^2+\chi^2}\,\mathrm{e}^{kL}/\Phi \qquad (3.31)$$

$$i/e=(k^2+\chi^2-\mu^2k^2)(\mathrm{e}^{\sqrt{k^2+\chi^2}L}-\mathrm{e}^{-\sqrt{k^2+\chi^2}L})\,\mathrm{e}^{2kL}/\Phi \qquad (3.32)$$

where

$$\Phi=(\sqrt{k^2+\chi^2}+\mu k)^2\,\mathrm{e}^{\sqrt{k^2+\chi^2}L}-(\sqrt{k^2+\chi^2}-\mu k)^2\,\mathrm{e}^{-\sqrt{k^2+\chi^2}L}. \qquad (3.33)$$

In (3.31), (3.32) and (3.33), suffixes m and n are dropped for the sake of simplicity. Coefficients C and D are also readily obtained although they are not given here explicitly.

When the time variation is periodic with period T, we put

$$p=i\omega,\quad (i=\sqrt{-1}) \qquad (3.34)$$

where ω is angular frequency defined by

$$\omega=2\pi/T\,. \qquad (3.35)$$

In that case, $\bar{e}/e$, i/e and Φ as defined respectively by (3.31), (3.32) and (3.33) are complex quantities. The attenuation factor or shielding ratio S of the slab, which may be defined by the ratio of the potential coefficient of the field beneath the slab, to that of the ambient external magnetic field is then given by

$$S=\mathrm{mod}\,\bar{e}/e\,. \qquad (3.36)$$

The phase difference between the magnetic fields in regions $z<-L$ and $z>0$ is given by

$$\alpha=\arg\bar{e}/e\,. \qquad (3.37)$$

For the purpose of numerically estimating typical values of S, we here assume $n=1$ and $m=0$. It is therefore seen that the magnetic field is sinusoidal in the x direction with a wavelength l and uniform in the y direction, so that we have

$$k = 2\pi / l. \tag{3.38}$$

Let us introduce an induction parameter which is defined by

$$\Omega = \sqrt{4\pi\mu\sigma\omega} L \tag{3.39}$$

where L is the thickness of the slab. It is easily seen that Ω is dimensionless.

From (3.26), we thus obtain

$$\chi^2 = i\,(\Omega/L)^2. \tag{3.40}$$

We further define λ as

$$\lambda = \chi/k = \frac{1+i}{\sqrt{2}}\,\Omega/kL. \tag{3.41}$$

With the aid of these relations, (3.31), (3.32) and (3.33) can be rewritten as

$$\bar{e}/e = 4\mu\sqrt{1+\lambda^2}\,e^{kL}/\Phi' \tag{3.42}$$

$$i/e = (1+\lambda^2-\mu^2)(e^{\sqrt{1+\lambda^2}kL} - e^{-\sqrt{1+\lambda^2}kL})\,e^{2kL}/\Phi' \tag{3.43}$$

$$\Phi' = (\sqrt{1+\lambda^2}+\mu)^2\,e^{\sqrt{1+\lambda^2}kL} - (\sqrt{1+\lambda^2}-\mu)^2\,e^{-\sqrt{1+\lambda^2}kL}. \tag{3.44}$$

When $\Omega=0$, we see that the amplitude ratio $\bar{e}/e$ becomes

$$\bar{e}/e = \frac{4\mu e^{kL}}{(1+\mu)^2\,e^{kL} - (1-\mu)^2\,e^{-kL}} \tag{3.45}$$

which is easily obtained when the slab is non-conducting (see Appendix 1). This is the shielding ratio for a static magnetic field due to a slab of which the magnetic permeability is μ.

It is somewhat tedious to calculate $\bar{e}/e$ and i/e for many combinations of μ, Ω or σ and kL from (3.42) and (3.43). Figure 3.2 shows how mod $\bar{e}/e$ and arg $\bar{e}/e$ change as Ω changes when $kL=0.1$ and $\mu=1$ are assumed. No graphs of a similar kind for other values of μ are shown here although there is no difficulty in calculating them.

It is demonstrated in the figure that the slab is almost transparent for variations having an induction parameter smaller than 0.1. On the other hand, the slab can shield the region beneath it from the external magnetic field almost perfectly when $\Omega \geqq 2$. The shielding ratio or amplitude ratio changes from 1 to 0 mainly over the range of Ω from 0.1 to 3. It is also noticeable that the phase difference between the inducing and induced fields changes from 0 to $\pi/2$ over that range of Ω. Changes in amplitude ratio and phase difference outside that range are small.

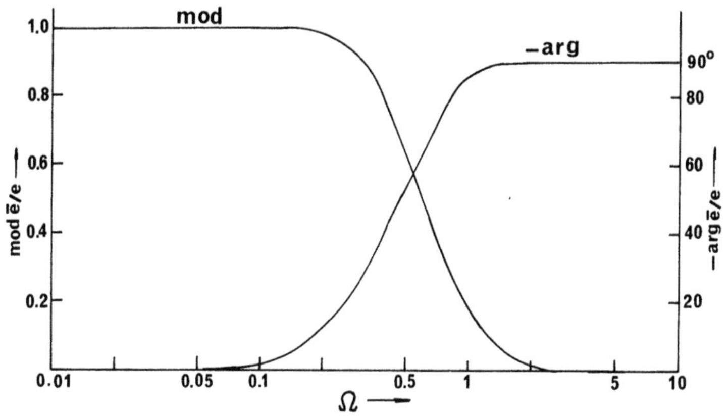

FIG. 3.2 Changes in mod $\overline{e}/e$ and arg $\overline{e}/e$ with induction parameter Ω.

In order to examine the magnetic field produced by the electric current induced in the slab, or the electromagnetic response of the slab, $(i/e)\,e^{-2kL}$ is calculated. Its modulus and argument are shown in Fig. 3.3 in which we see that the induced field becomes appreciably large when Ω exceeds 0.2. The change in phase difference becomes large for that range of Ω, too.

What we learned in this section can be applied to investigating the shielding effect of the earth's mantle if the sphericity of the earth is

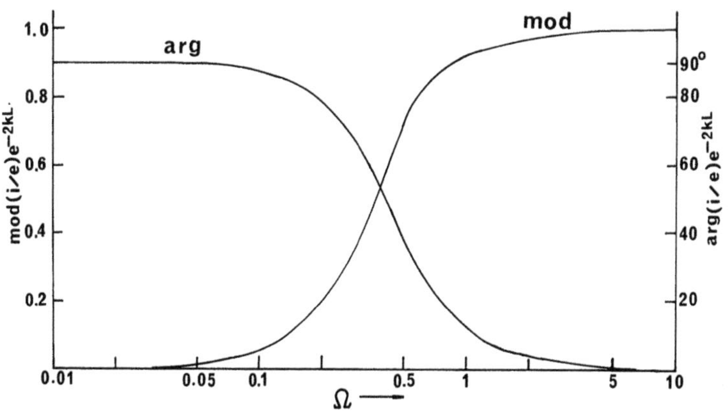

FIG. 3.3 Electromagnetic response of the slab, or modulus and argument of $(i/e)e^{-2kL}$, as a function of induction parameter Ω.

ignored. The geomagnetic secular variations having a period of a few to several hundred or thousand years are believed to be caused by some magneto-hydrodynamic processes in the fluid core beneath the mantle. As the earth's mantle is conducting (e.g., RIKITAKE, 1966), the secular variations should be subject to some attenuation while they diffuse through the mantle, so that we, who are living on the earth's surface, are actually observing variations which are affected by the mantle conductivity.

Figure 3.4 shows how the attenuation factor (A.F.) or amplitude ratio of the variation at the top of the mantle to that at the bottom of the mantle changes as the period of variation changes taking the mantle conductivity as a parameter. In the estimation, it is assumed that $L=3000$ km and $kL=0.1$.

Since it is believed that the conductivity of the lower mantle is of the order of 10 S/m ($=10\ \text{ohm}^{-1}\text{m}^{-1}=10^{-10}$ e.m.u.) (RIKITAKE, 1973), it seems likely that geomagnetic secular variations having a period of 100 years or shorter are strongly affected by the conducting mantle.

If we think of a copper slab of which the resistivity and thickness amount to 10^{-8} ohm m and 1 mm respectively, the induction parameter Ω becomes 1 for a frequency of about 2500 Hz on the condition that $kL=2\pi L/l=0.1$. Hence, we see that the slab strongly shields an electromagnetic variation having a frequency of 2500 Hz or over.

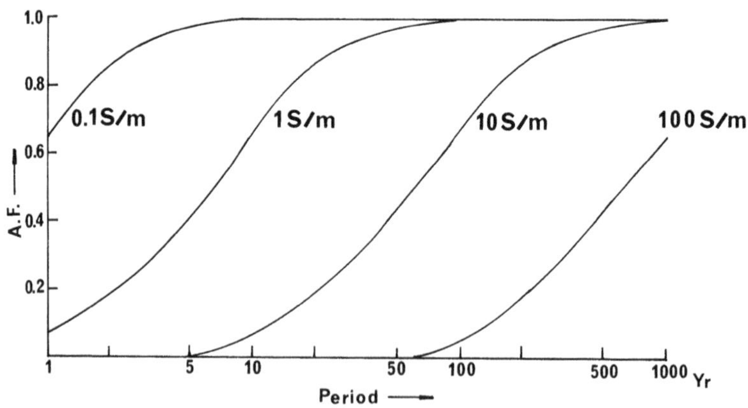

FIG. 3.4 Attenuation factor (A.F.) of geomagnetic secular variation as a function of period.

3.2.1 Line of magnetic dipoles over a uniform slab

In this subsection, electromagnetic induction by an oscillating dipole over a plane slab conductor will be studied. As the problem is two-dimensional, we actually deal with a line of time-dependent dipoles which is shown in Fig. 3.5.

The magnetic potential due to the line of dipoles at $z=-h, x=0$ is given as

$$w_0=M\frac{\cos\theta}{r} \tag{3.46}$$

where r and θ are defined as shown in Fig. 3.5, while M denotes the magnetic moment of the dipole.

(3.46) can be rewritten as

$$w_0=M\int_0^\infty e^{-(z+h)k}\sin kx\,dk \quad (0>z-h). \tag{3.47}$$

The magnetic potentials for $z<-L$ and $z>0$ and the vector potential for $-L<z<0$ as given in (3.19), (3.22) and (3.27) can be written in generalized forms such as

$$W=\int_0^\infty (Me^{-k(z+h)}+i(k)\,e^{kz})\sin kx\,dk \quad (z<-L), \tag{3.48}$$

$$\overline{W}=\int_0^\infty \bar{e}(k)\,e^{-kz}\sin kx\,dk \quad (z>0), \tag{3.49}$$

and

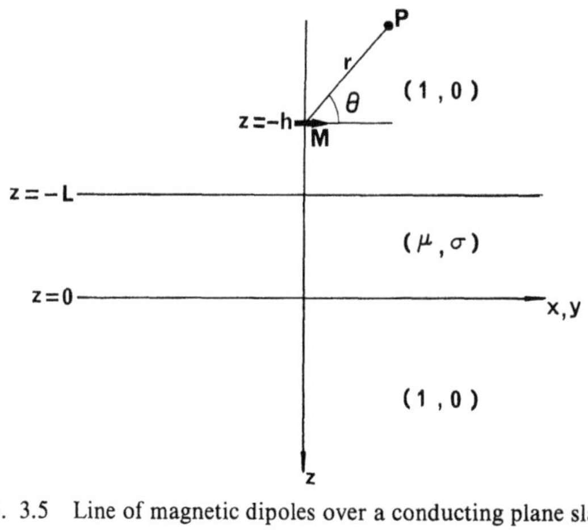

FIG. 3.5 Line of magnetic dipoles over a conducting plane slab.

$$A_y=\int_0^{\infty}(C(k)\,\mathrm{e}^{-\sqrt{k^2+\chi^2}z}+D(k)\,\mathrm{e}^{\sqrt{k^2+\chi^2}z})\sin kx\,\mathrm{d}k\quad(-L<z<0),\tag{3.50}$$

in which χ is defined by (3.26) together with (3.34) for a periodic variation. It is clear that A_x and A_z are zero in this case.

From the boundary conditions as given in (3.30), we obtain

$$\overline{W}=4\mu M\int_0^{\infty}\Phi^{-1}k\sqrt{k^2+\chi^2}\mathrm{e}^{k(L-z-h)}\sin kx\,\mathrm{d}k\tag{3.51}$$

$$W=w_0+M\int_0^{\infty}\Phi^{-1}(k^2+\chi^2-\mu^2k^2)(\mathrm{e}^{\sqrt{k^2+\chi^2}L}-\mathrm{e}^{-\sqrt{k^2+\chi^2}L})\cdot\mathrm{e}^{k(2L+z+h)}\sin kx\,\mathrm{d}k\tag{3.52}$$

where Φ is given by (3.33).

In spite of the success of obtaining analytical expressions for the magnetic potentials as obtained in (3.51) and (3.52), it is no easy matter to evaluate the integrals involved and to obtain the distribution of the magnetic field. It is certain, however, that the inducing dipole field is distorted by the induced filed, so that the magnetic fields above and below the slab are no longer of dipole type.

When the thickness of the slab is large, we see that

$$\lim_{L\to\infty}\Phi^{-1}(k^2+\chi^2-\mu^2k^2)(\mathrm{e}^{\sqrt{k^2+\chi^2}L}-\mathrm{e}^{-\sqrt{k^2+\chi^2}L})=\frac{\sqrt{k^2+\chi^2}-\mu k}{\sqrt{k^2+\chi^2}+\mu k}.$$

When $\mu=1$ and σ is large, we may perform an expansion such as

$$\frac{\sqrt{k^2+\chi^2}-k}{\sqrt{k^2+\chi^2}+k}=1-2\chi^{-1}k+2(\chi^{-1}k)^2-(\chi^{-1}k)^3+\cdots.$$

We, therefore, can rewrite (3.52) as

$$\begin{aligned}W=w_0&+M\int_0^{\infty}\mathrm{e}^{2kL}\mathrm{e}^{(z-h)k}\sin kx\,\mathrm{d}k\\&-2M\chi^{-1}\int_0^{\infty}\mathrm{e}^{2kL}\mathrm{e}^{(z-h)k}\sin kx\,(k\,\mathrm{d}k)\\&+2M\chi^{-2}\int_0^{\infty}\mathrm{e}^{2kL}\mathrm{e}^{(z-h)k}\sin kx\,(k^2\,\mathrm{d}k)\\&\cdots\cdots\\&\cdots\cdots.\end{aligned}\tag{3.53}$$

Let us then take polar coordinates $(r_1,\ \theta_1)$ for which the origin is located at $z=h-2L$ on the z axis as shown in Fig. 3.6. In that case, we can prove the following relations:

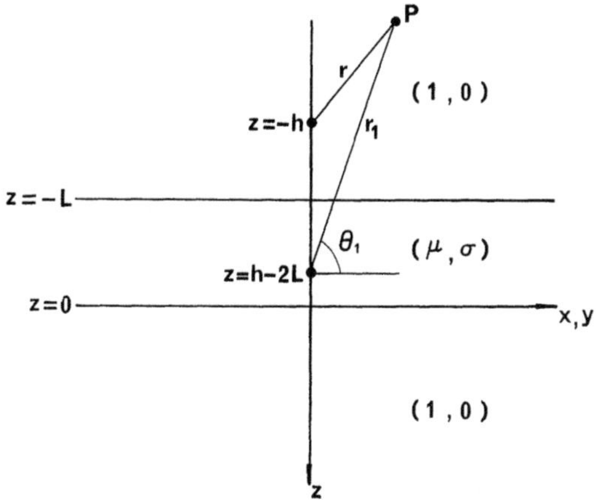

FIG. 3.6 Line of magnetic dipoles and its mirror projection.

$$\begin{aligned}
&\int_0^\infty e^{2kL}e^{(z-h)k} \sin kx \, d\,k = r_1^{-1} \cos \theta_1 \\
&\int_0^\infty e^{2kL}e^{(z-h)k} \sin kx \, (k \, d\,k) = r_1^{-2} \sin 2\theta_1 \\
&\int_0^\infty e^{2kL}e^{(z-h)k} \sin kx \, (k^2 \, d\,k) = -2r_1^{-3} \cos 3\theta_1 \\
&\cdots\cdots \\
&\cdots\cdots
\end{aligned} \tag{3.54}$$

as shown by RIKITAKE (1966).

Putting (3.54) into (3.53), we obtain

$$\begin{aligned} W = w_0 + Mr_1^{-1} \sin 2\theta_1 - 2M\chi^{-1}r_1^{-2} \sin 2\theta_1 \\ -4M\chi^{-2}r_1^{-3} \sin 3\theta_1 \cdots\cdots . \end{aligned} \tag{3.55}$$

We see, therefore, that the induced potential is the sum of magnetic potentials of a line of dipoles, a line of quadrupoles and so on placed at the mirror point of the original line of dipoles with respect to the upper surface of the conducting slab. Such a feature of the induced field is exactly the same as that in the case of electromagnetic induction in a semi-infinite uniform medium. If $\sigma \rightarrow \infty$, the induced field is simply due to a line of dipoles the strength of which is the same as that of the inducing dipoles as has been shown by RIKITAKE (1966).

3.3 Spherical Shell

Electromagnetic induction in a sphere has been studied by CHAPMAN (1919), CHAPMAN and PRICE (1930), LAHIRI and PRICE (1939) and others mostly in relation to electromagnetic induction within the earth. Summaries of these studies are given in CHAPMAN and BARTELS (1940) and RIKITAKE (1966). SUGIURA and NAGATA (1949) discussed the shielding effect of the ionosphere on the basis of electromagnetic induction in a uniform spherical shell.

The theory of electromagnetic induction in a uniform spherical shell by a time-dependent magnetic field arising from outside or inside the shell will be summarized in the following subsections.

3.3.1 Conductors with spherical symmetry

Let us assume that

$$\sigma=\sigma(\rho) \tag{3.56}$$

where ρ measures distance from the shell's center in units of the shell's outer radius R_1, and is defined by

$$\rho=R/R_1 \tag{3.57}$$

in the spherical polar coordinate system (R, θ, ϕ).

LAHIRI and PRICE (1939) have shown that the elementary solution of (3.17) is of the form

$$\boldsymbol{A}=\boldsymbol{r}\wedge \operatorname{grad} u \tag{3.58}$$

where u satisfies

$$\nabla^2 u=4\pi\sigma\mu\partial u/\partial t. \tag{3.59}$$

It is readily seen that (3.58) satisfies (3.15). Since $\boldsymbol{A}$ has no radial component, the electric currents flow perpendicular to any radius, so that the condition $\boldsymbol{i}\perp\operatorname{grad}\sigma$ is satisfied. A vector of this type is usually called toroidal. Although there is a vector of another type called poloidal which also satisfies div $\boldsymbol{A}=0$, it can be shown that a vector potential of the latter type does not result in a magnetic field outside the conductor, and hence does not enter into the problem of electromagnetic induction by an external magnetic field.

A typical solution of (3.59) is given by

$$u_n=R_1 f_n(p, \rho)S_n \tag{3.60}$$

where S_n is a surface harmonic of degree n, and f_n is a solution of the radial differential equation

$$(\mathrm{d}/\mathrm{d}\,\rho)(\rho^2\,\mathrm{d}\,f_n/\mathrm{d}\,\rho)=[n\,(n+1)+4\pi\mu R_1^2\rho^2\sigma\,(\rho)\,p]\,f_n \qquad (3.61)$$

in which $\partial/\partial t$ has been replaced by an algebraic operator p. A typical solution of (3.58) therefore becomes

$$\boldsymbol{A}_n=R_1\,(\boldsymbol{r}\wedge\mathrm{grad}\ S_n)\,f_n\,(p,\,\rho). \qquad (3.62)$$

The components of the magnetic flux density are then given as

$$\begin{aligned} B_R&=-\rho^{-1}n\,(n+1)\,f_nS_n\\ B_\theta&=-\rho^{-1}\,\mathrm{d}\,(\rho f_n)/\mathrm{d}\,\rho\cdot\partial S_n/\partial\theta\\ B_\phi&=-\rho^{-1}\,\mathrm{d}\,(\rho f_n)/\mathrm{d}\,\rho\cdot\partial S_n/\sin\theta\partial\phi. \end{aligned} \qquad (3.63)$$

3.3.2 Electromagnetic induction by a magnetic field arising from outside a spherical shell

Magnetic potentials W_0 and W_2 can be defined in the non-conducting regions $R>R_1$ and $R_2>R$, respectively, where R_2 is the inner radius of the spherical shell.

Since W_0 and W_2 satisfy

$$\nabla^2W_0=0 \qquad (R>R_1) \qquad (3.64)$$

and

$$\nabla^2W_2=0 \qquad (R_2>R), \qquad (3.65)$$

typical solutions for the respective regions are given as

$$W_{0,\,n}=R_1\,(e_n\rho^n+i_n\rho^{-n-1})\,S_n \qquad (R>R_1) \qquad (3.66)$$

and

$$W_{2,\,n}=R_1\bar{e}_n\rho^nS_n \qquad (R_2>R). \qquad (3.67)$$

It is assumed here that no sources exist in the region surrounded by the shell. From (3.66) and (3.67), the magnetic field components are obtained as

$$\begin{aligned} H_{0,R}&=-[ne_n\rho^{n-1}-(n+1)\,i_n\rho^{-n-2}]S_n\\ H_{0,\theta}&=-(e_n\rho^{n-1}+i_n\rho^{-n-2})\,\partial S_n/\partial\theta\\ H_{0,\phi}&=-(e_n\rho^{n-1}+i_n\rho^{-n-2})\,\partial S_n/\sin\theta\partial\phi \end{aligned} \qquad (R>R_1) \qquad (3.68)$$

$$\begin{aligned} H_{2,R}&=-n\bar{e}_n\rho^{n-1}S_n\\ H_{2,\theta}&=-\bar{e}_n\rho^{n-1}\partial S_n/\partial\theta\\ H_{2,\phi}&=-\bar{e}_n\rho^{n-1}\partial S_n/\sin\theta\partial\phi \end{aligned} \qquad (R_2>R). \qquad (3.69)$$

The continuity of tangential magnetic field and normal magnetic flux density at the outer and inner boundaries of the shell implies the following relations:

$$\begin{aligned} &ne_n - (n+1)\, i_n = n\,(n+1)\, f_{n,\,\rho=1} \\ &e_n + i_n = \mu^{-1}\,[\mathrm{d}\,(\rho f_n)/\mathrm{d}\rho]_{\rho=1} \\ &n\,(n+1)(\rho^{-1} f_n)_{\rho=q} = n\bar{e}_n q^{n-1} \\ &\mu^{-1}\,[\rho^{-1}\,\mathrm{d}\,(\rho f_n)/\mathrm{d}\,\rho]_{\rho=q} = \bar{e}_n q^{n-1}, \end{aligned} \tag{3.70}$$

where

$$q = R_2/R_1\,. \tag{3.71}$$

In order to determine the relation between e_n, i_n and $\bar{e}_n$ on the basis of (3.70), it is necessary to know the functional form of f_n, which is a solution of (3.61).

When σ is a constant, (3.61) reduces to

$$\mathrm{d}^2 f_n/\mathrm{d}\,\rho^2 + 2\rho^{-1}\,\mathrm{d}\,f_n/\mathrm{d}\,\rho - [n\,(n+1)\,\rho^{-2} + k^2 R_1^2]\, f_n = 0 \tag{3.72}$$

by putting

$$k^2 = 4\pi\mu\sigma p\,. \tag{3.73}$$

It is easily seen that particular solutions of (3.72) are

$$\rho^{-1/2} I_\nu(kR_1\rho) \quad \text{and} \quad \rho^{-1/2} K_\nu(kR_1\rho)$$

where I_ν and K_ν are modified Bessel functions, with ν given by

$$\nu = n + 1/2\,. \tag{3.74}$$

We hence obtain

$$f_n = \rho^{-1/2}[A_n I_\nu(z) + B_n K_\nu(z)] \tag{3.75}$$

in which

$$z = kR_1\,\rho\,. \tag{3.76}$$

Putting (3.75), which is the general form of f_n, into (3.70), the boundary conditions can be rewritten

$$\begin{aligned} &ne_n - (n+1)\, i_n = n\,(n+1)[A_n I_\nu(z_1) + B_n K_\nu\,(z_1)] \\ &2\mu\,(e_n + i_n) = A_n\,[I_\nu\,(z_1) + 2z_1 I_\nu'\,(z_1)] \\ &\qquad + B_n\,[K_\nu(z_1) + 2z_1 K_\nu'\,(z_1)] \\ &A_n\,[I_\nu\,(z_q) + 2z_q I_\nu'\,(z_q)] + B_n\,[K_\nu\,(z_q) + 2z_q K_\nu'\,(z_q)] \\ &\qquad = 2\mu q^{n+1/2}\bar{e}_n \\ &(n+1)[A_n I_\nu(z_q) + B_n K_\nu(z_q)] = q^{n+1/2}\bar{e}_n \end{aligned} \tag{3.77}$$

where I' and K' denote dI/dz and dK/dz, respectively. z_1 and z_q are defined by $z_1=kR_1$ and $z_q=kR_1q$.

Solving (3.77), we obtain

$$\bar{e}_n/e_n=\Psi^{-1}nq^{-\nu}(2n+1) \tag{3.78}$$

$$\begin{aligned} i_n/e_n=\Psi^{-1}\frac{n}{n+1}\Big[&\Big\{\Big(-(n+1)\,\mu+\frac{1}{2}\Big)\,I_\nu(z_1)+z_1I_\nu{}'(z_1)\Big\} \\ &\cdot\Big\{(n+1)\,\mu K_\nu(z_q)-\frac{1}{2}\,(K_\nu(z_q)+2z_qK_\nu{}'(z_q)\Big\} \\ &-\Big\{\Big(-(n+1)\,\mu+\frac{1}{2}\Big)\,K_\nu(z_1)+z_1K_\nu{}'(z_1)\Big\} \\ &\cdot\Big\{(n+1)\,\mu I_\nu(z_q)-\frac{1}{2}\,(I_\nu(z_q)+2z_qI_\nu{}'(z_q)\Big\}\Big] \end{aligned} \tag{3.79}$$

$$A_n/\bar{e}_n=\frac{q^\nu}{n+1}\Big[(n+1)\,\mu K_\nu(z_1)-\frac{1}{2}\,(K_\nu(z_1)+2z_1K_\nu{}'(z_1))\Big] \tag{3.80}$$

$$B_n/\bar{e}_n=-\frac{q^\nu}{n+1}\Big[(n+1)\,\mu I_\nu(z_1)-\frac{1}{2}\,(I_\nu(z_1)+2z_1I_\nu{}'(z_1))\Big] \tag{3.81}$$

where

$$\begin{aligned} \Psi=&\Big\{\Big(n\mu+\frac{1}{2}\Big)I_\nu(z_1)+z_1I_\nu{}'(z_1)\Big\} \\ &\cdot\Big\{(n+1)\,\mu K_\nu(z_q)-\frac{1}{2}\,(K_\nu(z_q)+2z_qK_\nu{}'(z_q))\Big\} \\ &-\Big\{\Big(n\mu+\frac{1}{2}\Big)\,K_\nu(z_1)+z_1K_\nu{}'(z_1)\Big\} \\ &\cdot\Big\{(n+1)\,\mu I_\nu(z_q)-\frac{1}{2}\,(I_\nu(z_q)+2z_qI_\nu{}'(z_q))\Big\}. \end{aligned} \tag{3.82}$$

Let us consider a very thick shell, and investigate the case when $q\to0$. Then $I_\nu(z_q)\to0$ and $qI'_\nu(z_q)\to0$, so that (3.79) simplifies to

$$i_n/e_n=\frac{n}{n+1}\,\frac{\Big\{-(n+1)\,\mu+\frac{1}{2}\Big\}\,I_\nu(z_1)+z_1I_\nu{}'(z_1)}{\Big(n\mu+\frac{1}{2}\Big)\,I_\nu(z_1)+z_1I_\nu{}'(z_1)}. \tag{3.83}$$

With the aid of the formula

$$\frac{\nu}{z}\,I_\nu(z)+I_\nu{}'(z)=I_{\nu-1}(z) \tag{3.84}$$

for modified Bessel functions, (3.83) becomes

$$i_n/e_n=\frac{n}{n+1}\Big[1-\frac{(2n+1)\,\mu I_\nu(z_1)}{n(\mu-1)\,I_\nu(z_1)+z_1I_{\nu-1}(z_1)}\Big] \tag{3.85}$$

which may be further rewritten as

$$i_n/e_n = \frac{n}{n+1}\left[1-\mu\left\{\frac{kR_1}{2n+1}\frac{I_{n-1/2}(kR_1)}{I_{n+1/2}(kR_1)}+\frac{n(\mu-1)}{2n+1}\right\}^{-1}\right]. \quad (3.86)$$

i_n/e_n given by (3.86) agrees with the amplitude ratio of the coefficient of the internal magnetic potential to that of the external magnetic potential when electromagnetic induction in a uniform sphere of radius R_1 is considered (e.g., RIKITAKE, 1966, p. 130).

Next, let us study an extreme case when $\sigma\to 0$. From the definitions of modified Bessel functions, we have

$$\begin{aligned}
&\lim_{z\to 0} I_\nu(z) \approx \frac{(z/2)^\nu}{\nu\Gamma(\nu)}\\
&\lim_{z\to 0} I_\nu'(z) \approx \frac{(z/2)^{\nu-1}}{2\Gamma(\nu)}\\
&\lim_{z\to 0} K_\nu(z) \approx \frac{1}{2}\Gamma(\nu)(z/2)^{-\nu}\\
&\lim_{z\to 0} K_\nu'(z) \approx -\frac{1}{4}\nu\Gamma(\nu)(z/2)^{-\nu-1}
\end{aligned} \quad (3.87)$$

where $\Gamma(\nu)$ is the gamma function.

Substituting (3.87) into (3.82), we obtain

$$\begin{aligned}
\Psi = &\frac{q^{-n-\frac{1}{2}}}{2n+1}[n(n+1)\mu^2+(2n^2+2n+1)\mu+n(n+1)]\\
&-\frac{q^{n+\frac{1}{2}}}{2n+1}[n(n+1)\mu^2-(2n^2+2n)\mu+n(n+1)]
\end{aligned} \quad (3.88)$$

so that we arrive at

$$\frac{\bar{e}_n}{e} = \frac{1}{1-\dfrac{n(n+1)}{(2n+1)^2}\dfrac{(\mu-1)^2}{\mu}(1-q^{2n+1})} \quad (3.89)$$

which is the shielding ratio of the shell when the field is static. If we put $n=1$, (3.89) agrees with (2.17) which was obtained for a uniform magnetic field.

We thus see in this subsection that the expressions (3.78) and (3.79) reduce to previously-obtained results in the above two extreme cases.

3.3.3 *Shielding effect to an external field*

As it would be laborious to estimate shielding for the various degrees of spherical harmonics which represent inducing magnetic fields, let us take, as one example, $n=1$ which corresponds to a uniform inducing field. $\mu=1$ is also assumed throughout the following derivation.

In that case, (3.78), (3.79) and (3.82) reduce to

$$\bar{e}_1/e_1=3\Psi^{-1}q^{-3/2} \tag{3.90}$$

$$i_1/e_1=\frac{1}{2}\Psi^{-1}\left[I_{\frac{5}{2}}(z_1)\,K_{\frac{5}{2}}(z_q)-K_{\frac{5}{2}}(z_1)\,I_{\frac{5}{2}}(z_q)\right]z_1z_q \tag{3.91}$$

$$\Psi=z_1z_q\left[I_{\frac{1}{2}}(z_1)\,K_{\frac{5}{2}}(z_q)-K_{\frac{1}{2}}(z_1)\,I_{\frac{5}{2}}(z_q)\right]. \tag{3.92}$$

For a purely periodic variation of angular frequency ω, z_1 and z_q are given as

$$z_1=\Omega\sqrt{i},\quad z_q=q\Omega\sqrt{i}\quad (i=\sqrt{-1})$$

where Ω is the induction parameter which is given by

$$\Omega=\sqrt{4\pi\sigma\omega}R_1$$

in a similar manner to (3.39).

With the aid of the above equations, the real and imaginary parts of $\bar{e}_1/e_1$ for various Ω may be determined. Tables 3.1 and 3.2 show the calculations for $q=0.8$ and 0.4. The shielding ratio S_i, which is given by mod $\bar{e}_1/e_1$ is calculated along with arg $\bar{e}_1/e_1$ as shown in Figs. 3.7 and 3.8 for $q=0.8$ and 0.4, respectively. It is evident from the figures that S_i changes very steeply between $\Omega=2$ and 10. The shell is almost transparent for variations having an induction parameter small than 1. Conversely, the region inside the shell is shielded almost perfectly from the external magnetic field for $\Omega\geqq 50$.

Let us imagine a hypothetical earth's mantle having an electrical conductivity amounting to 10 S/m and a thickness of $0.4a$, where a is the earth's radius ($a=6400$ km). In that case the induction parameter is $\Omega=4.05$, 1.28 and 0.405 for geomagnetic variations of periods 1, 10 and 100 years, respectively. We therefore see that a geomagnetic variation having a period of 1 year or shorter is subject to considerable shielding. On the contrary, a variation of period longer than 100 years can penetrate through such a mantle without being influenced by electromagnetic shielding. As will be seen in Subsection 3.3.4, the conclusion may be applied to a geomagnetic secular variation arising from a source inside the mantle.

Let us now assume that the shell is composed of copper having a resistivity amounting to 10^{-8} ohm m, that the ratio of the outer radius to

TABLE 3.1 Real and imaginary parts of $\bar{e}_1/e_1$ for q=0.8.

Ω	Re($\tilde{e}_1/e_1$)	Im($\tilde{e}_1/e_1$)
0.5	0.9998	−0.0150
1.0	0.9967	−0.0598
1.5	0.9837	−0.1330
2.0	0.9502	−0.2292
2.5	0.8857	−0.3365
3.0	0.7863	−0.4363
3.5	0.6595	−0.5100
4.0	0.5217	−0.5471
4.5	0.3905	−0.5489
5.0	0.2773	−0.5246
5.5	0.1860	−0.4851
6.0	0.1154	−0.4391
7.0	0.0226	−0.3476
8.0	−0.0282	−0.2694
9.0	−0.0552	−0.2065
10.0	−0.0685	−0.1566
20.0	−0.0229	0.0081
30.0	0.0009	0.0040
40.0	0.0008	−0.0001
50.0	0.0000	−0.0002

the inner one is 0.8, and that the outer radius is 10 cm, the induction parameter is then calculated as 0.889, 2.81, 8.89 and 28.1 for alternating magnetic fields of frequency 0.1, 1, 10 and 100 Hz, respectively. It is then obvious that a field, for which the frequency is a few tens of Hz, undergoes appreciable shielding, that is the shielding ratio becomes less than 0.1. Meanwhile, a field having a frequency of 0.1 Hz or less cannot be shielded by the spherical shell. Such a conclusion is important for making the shielding effective against magnetic fields arising from commercial electric power lines. Since the induction parameter is proportional to the radius of the shell, it is recommended for effective shielding to make use of a shell as large as possible.

No numerical results for the electromagnetic response of a spherical shell, or i_1/e_1 as given in (3.91), are presented here for the sake of simplicity. It is easy to perform the computer calculation, however.

TABLE 3.2 Real and imaginary parts of $\bar{e}_1/e_1$ for q=0.4.

Ω	Re($\bar{e}_1/e_1$)	Im($\bar{e}_1/e_1$)
0.5	0.9991	−0.0350
1.0	0.9853	−0.1385
1.5	0.9283	−0.2989
2.0	0.7945	−0.4777
2.5	0.5814	−0.6119
3.0	0.3346	−0.6538
3.5	0.1145	−0.6056
4.0	−0.0455	−0.5030
4.5	−0.1428	−0.3825
5.0	−0.1895	−0.2673
5.5	−0.2001	−0.1687
6.0	−0.1875	−0.0909
7.0	−0.1311	0.0049
8.0	−0.0714	0.0405
9.0	−0.0283	0.0427
10.0	−0.0039	0.0316
20.0	−0.0003	−0.0001
30.0	0.0000	0.0000
40.0	0.0000	0.0000
50.0	0.0000	0.0000

3.3.4 *Electromagnetic induction by a magnetic field arising from inside a spherical shell*

In the case of electromagnetic induction by a magnetic field arising from inside a spherical shell, we see that

$$W_{0,n}=R_1 i_n \rho^{-n-1} S_n \qquad (R>R_1) \qquad (3.93)$$

and

$$W_{2,n}=R_1\left(\bar{e}_n\rho^n+\bar{i}_n\rho^{-n-1}\right)S_n \qquad (R_2>R) \qquad (3.94)$$

are the typical solutions of (3.64) and (3.65), respectively. $\bar{i}_n$ represents the coefficient of magnetic potential arising from inside the shell, while we have no term representing the magnetic field arising from outside the shell in (3.93).

The magnetic field components are then given as

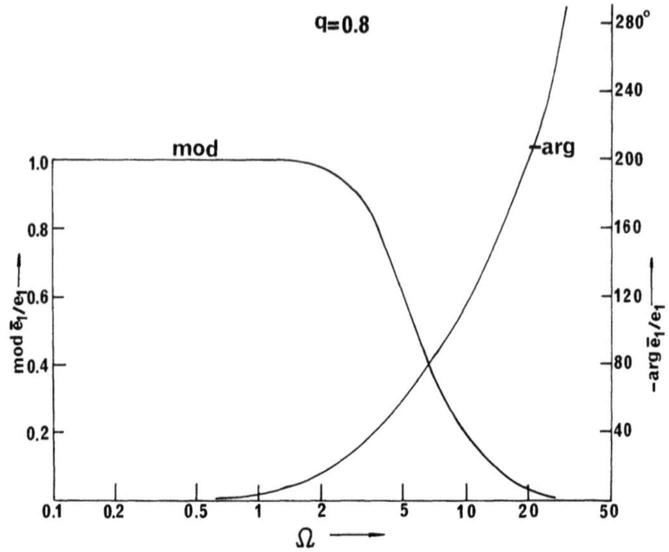

FIG. 3.7 mod $\bar{e}_1/e_1$ and arg $\bar{e}_1/e_1$ as functions of induction parameter Ω for a spherical shell of q=0.8.

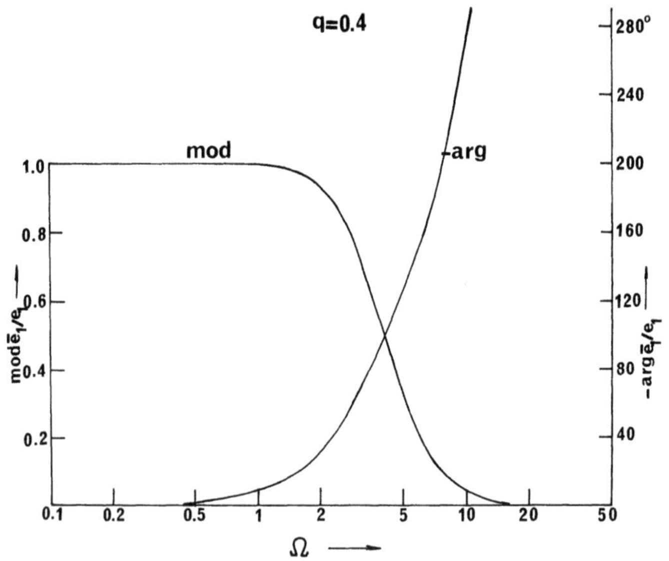

FIG. 3.8 mod $\bar{e}_1/e_1$ and arg $\bar{e}_1/e_1$ as functions of induction parameter Ω for a spherical shell of q=0.4.

$$
\begin{aligned}
H_{0,R} &= (n+1)\, i_n \rho^{-n-1} S_n \\
H_{0,\theta} &= -i_n \rho^{-n-2} \partial S_n/\partial\theta \qquad (R>R_1) \qquad (3.95) \\
H_{0,\phi} &= -i_n \rho^{-n-2} \partial S_n/\sin\theta\partial\phi
\end{aligned}
$$

and

$$
\begin{aligned}
H_{2,R} &= -[n\bar{e}_n \rho^{n-1} - (n+1)\, \bar{i}_n \rho^{-n-2}] S_n \\
H_{2,\theta} &= -(\bar{e}_n \rho^{n-1} + \bar{i}_n \rho^{-n-2})\, \partial S_n/\partial\theta \qquad (R_2>R) \qquad (3.96) \\
H_{2,\phi} &= -(\bar{e}_n \rho^{n-1} + \bar{i}_n \rho^{-n-2})\, \partial S_n/\sin\theta\partial\phi
\end{aligned}
$$

in the respective regions.

A vector potential, of which the typical term is given by (3.62), can be defined in the conducting shell, so that the components of magnetic flux density in the shell are given by (3.63).

The boundary conditions which are the same as adopted in Subsection 3.3.2 then imply

$$
\begin{aligned}
&-(n+1)\, i_n = n\,(n+1)\, f_{n,\rho=1} \\
&i_n = \mu^{-1}\, [\mathrm{d}\,(\rho f_n)/\mathrm{d}\rho]_{\rho=1} \\
&n\,(n+1)(\rho^{-1} f_n)_{\rho=q} = n\bar{e}_n q^{n-1} - (n+1)\, \bar{i}_n q^{-n-2} \\
&\mu^{-1}\, [\rho^{-1}\,\mathrm{d}\,(\rho f_n)/\mathrm{d}\,\rho]_{\rho=q} = \bar{e}_n q^{n-1} + \bar{i}_n q^{-n-2}\,.
\end{aligned} \qquad (3.97)
$$

As f_n is given by (3.75) along with (3.76), (3.97) can be rewritten as

$$
\begin{aligned}
&-i_n = n\,[A_n I_\nu(z_1) + B_n K_\nu(z_1)] \\
&2\mu i_n = A_n\,[I_\nu(z_1) + 2 z_1 I_\nu'(z_1)] + B_n\,[K_\nu(z_1) + 2 z_1 K_\nu'(z_1)] \\
&n\,(n+1)[A_n I_\nu(z_q) + B_n K_\nu(z_q)] = n\bar{e}_n q^\nu - (n+1)\bar{i}_n q^{-\nu} \\
&A_n\,[I_\nu(z_q) + 2 z_q I_\nu'(z_q)] + B_n\,[K_\nu(z_q) + 2 z_q K_\nu'(z_q)] \\
&\qquad = 2\mu(\bar{e}_n q^\nu + \bar{i}_n q^{-\nu})
\end{aligned} \qquad (3.98)
$$

where $\nu = n+1/2$, $z_1 = kR_1$ and $z_q = qkR_1$ as before.

Solving (3.98), we obtain

$$
A_n = \frac{1}{n}\left[\left(n\mu + \frac{1}{2}\right) K_\nu(z_1) + z_1 K_\nu'(z_1)\right] i_n \qquad (3.99)
$$

$$
B_n = -\frac{1}{n}\left[\left(n\mu + \frac{1}{2}\right) I_\nu(z_1) + z_1 I_\nu'(z_1)\right] i_n \qquad (3.100)
$$

$$
\begin{aligned}
\bar{i}_n = -\mu^{-1} \frac{n}{2n+1} q^\nu \Bigg[A_n &\left\{\left((n+1)\,\mu - \frac{1}{2}\right) I_\nu(z_q) - z_q I_\nu'(z_q)\right\} \\
+ B_n &\left\{\left((n+1)\,\mu - \frac{1}{2}\right) K_\nu(z_q) - z_q K_\nu'(z_q)\right\}\Bigg]
\end{aligned} \qquad (3.101)
$$

$$\bar{e}_n=\mu^{-1}\frac{n+1}{2n+1}q^{-\nu}\Big[A_n\Big\{\Big(n\mu+\frac{1}{2}\Big)I_\nu(z_q)+z_qI_\nu'(z_q)\Big\}$$
$$+B_n\Big\{\Big(n\mu+\frac{1}{2}\Big)K_\nu(z_q)+z_qK_\nu'(z_q)\Big\}\Big] \tag{3.102}$$

so that the ratio of i_n to $\bar{i}_n$, which is the shielding ratio, becomes

$$i_n/\bar{i}_n=\Phi^{-1}\mu q^{-\nu}(2n+1) \tag{3.103}$$

where

$$\Phi=\Big\{\Big(n\mu+\frac{1}{2}\Big)I_\nu(z_1)+z_1I_\nu'(z_1)\Big\}\Big\{\Big((n+1)\mu-\frac{1}{2}\Big)K_\nu(z_q)-z_qK_\nu'(z_q)\Big\}$$
$$-\Big\{\Big(n\mu+\frac{1}{2}\Big)K_\nu(z_1)+z_1K_\nu'(z_1)\Big\}$$
$$\cdot\Big\{\Big((n+1)\mu-\frac{1}{2}\Big)I_\nu(z_q)-z_qI_\nu'(z_q)\Big\}. \tag{3.104}$$

It is then obvious that

$$\Phi=\Psi \tag{3.105}$$

where Ψ has been obtained in (3.82).

When $\sigma\to 0$, (3.103) reduces to

$$i_n/\bar{i}_n=\frac{1}{1-\dfrac{n(n+1)}{(2n+1)^2}\dfrac{(\mu-1)^2}{\mu}(1-q^{2n+1})} \tag{3.106}$$

from (3.87) and (3.88). (3.106) agrees exactly with (3.89).

3.3.5 Shielding effect to an internal field

Assuming $\mu=1$ and $n=1$, (3.103) and (3.104) become

$$i_1/\bar{i}_1=3\Phi^{-1}q^{-3/2} \tag{3.107}$$

$$\Phi=z_1z_q\,[I_{\frac{1}{2}}(z_1)\,K_{\frac{5}{2}}(z_q)-K_{\frac{1}{2}}(z_1)\,I_{\frac{5}{2}}(z_q)]. \tag{3.108}$$

We, therefore, see that the degree of shielding to a dipole field arising from inside the shell is exactly the same as that to an external field which is given by (3.90) and (3.92). The shielding effect in this case can therefore be inferred from the results of Subsection 3.3.3.

3.4 Cylindrical Shell of Infinite Length

In this section, electromagnetic induction in a cylindrical shell having a circular cross section will be studied. The length of the shell is assumed to be infinite. We take cylindrical coordinates (r, θ, z), where the z axis coincides with the central axis.

The shell, in which the magnetic permeability is μ, occupies the region defined by $r_1>r>r_2$, while the regions defined by $r>r_1$ and $r_2>r$ are free space for which $\mu=1$.

3.4.1 Electromagnetic induction by a magnetic field arising from outside a cylindrical shell

In the free spaces, we can define magnetic potentials W_0 and W_2 of which the typical terms are given as

$$W_{0,n}=r^n(a_n\cos n\theta+b_n\sin n\theta)+r^{-n}(A_n\cos n\theta+B_n\sin n\theta) \qquad (r>r_1) \quad (3.109)$$

$$W_{2,n}=r^n(\bar{a}_n\cos n\theta+\bar{b}_n\sin n\theta) \qquad (r_2>r), \quad (3.110)$$

respectively. It is obvious that (3.109) and (3.110) satisfy $\nabla^2 W=0$.

We here assume that no magnetic field arises from sources in the region $r_2>r$, so that the following theory can be applicable to electromagnetic induction by a magnetic field arising from outside the shell.

In the shell, we define a vector potential $\boldsymbol{A}$, of which the z component is the only non-zero component. A_z should satisfy

$$\nabla^2 A_z=4\pi\sigma\mu\partial A_z/\partial t. \quad (3.111)$$

Putting

$$\partial/\partial t=p, \qquad k^2=4\pi\sigma\mu p \quad (3.112)$$

as before, a typical solution for A_z is found to be

$$A_{z,n}=I_n(kr)(C_n\cos n\theta+D_n\sin n\theta)+K_n(kr)(E_n\cos n\theta+F_n\sin n\theta) \qquad (r_2>r>r_1) \quad (3.113)$$

where I_n and K_n are modified Bessel functions.

From (3.109) and (3.110), the magnetic field components in free space are obtained as

$$\begin{aligned} H_{0,r}&=-nr^{n-1}(a_n\cos n\theta+b_n\sin n\theta)+nr^{-n-1}(A_n\cos n\theta+B_n\sin n\theta)\\ H_{0,\theta}&=nr^{n-1}(a_n\sin n\theta-b_n\cos n\theta)+nr^{-n-1}(A_n\sin n\theta-B_n\cos n\theta) \end{aligned} \qquad (r>r_1) \quad (3.114)$$

and

$$\begin{aligned} H_{2,r} &= -nr^{n-1}(\bar{a}_n \cos n\theta + \bar{b}_n \sin n\theta) \\ H_{2,\theta} &= nr^{n-1}(\bar{a}_n \sin n\theta - \bar{b}_n \cos n\theta) \end{aligned} \qquad (r_2 > r). \quad (3.115)$$

Meanwhile, the magnetic flux density components in the shell are obtained from (3.113) as

$$\begin{aligned} B_r &= -nr^{-1}I_n(kr)(C_n \sin n\theta - D_n \cos n\theta) \\ &\quad - nr^{-1}K_n(kr)(E_n \sin n\theta - F_n \cos n\theta) \\ B_\theta &= -kI_n'(kr)(C_n \cos n\theta + D_n \sin n\theta) \\ &\quad - kK_n'(kr)(E_n \cos n\theta + F_n \sin n\theta) \end{aligned} \qquad (r_1 > r > r_2) \quad (3.116)$$

where $I_n'(x) = \mathrm{d}I_n(x)/\mathrm{d}x$ and $K_n'(x) = \mathrm{d}K_n(x)/\mathrm{d}x$ as before.

The boundary conditions at $r = r_1$ imply

$$\begin{aligned} r_1^{n-1}a_n - r_1^{-n-1}A_n &= -r_1^{-1}[I_n(kr_1)D_n + K_n(kr_1)F_n] \\ n(r_1^{n-1}a_n + r_1^{-n-1}A_n) &= -\mu^{-1}k[I_n'(kr_1)D_n + K_n'(kr_1)F_n] \end{aligned} \qquad (3.117)$$

$$\begin{aligned} r_1^{n-1}b_n - r_1^{-n-1}B_n &= r_1^{-1}[I_n(kr_1)C_n + K_n(kr_1)E_n] \\ n(r_1^{n-1}b_n + r_1^{-n-1}B_n) &= \mu^{-1}k[I_n'(kr_1)C_n + K_n'(kr_1)E_n] \end{aligned} \qquad (3.118)$$

which reduce to

$$\begin{aligned} a_n &= -\frac{r_1^{-n}}{2n\mu}[(kr_1 I_n'(1) + n\mu I_n(1))D_n \\ &\qquad + (kr_1 K_n'(1) + n\mu K_n(1))F_n] \\ A_n &= -\frac{r_1^{n}}{2n\mu}[(kr_1 I_n'(1) - n\mu I_n(1))D_n \\ &\qquad + (kr_1 K_n'(1) - n\mu K_n(1))F_n] \end{aligned} \qquad (3.119)$$

$$\begin{aligned} b_n &= \frac{r_1^{-n}}{2n\mu}[(kr_1 I_n'(1) + n\mu I_n(1))C_n \\ &\qquad + (kr_1 K_n'(1) + n\mu K_n(1))E_n] \\ B_n &= \frac{r_1^{n}}{2n\mu}[(kr_1 I_n'(1) - n\mu I_n(1))C_n \\ &\qquad + (kr_1 K_n'(1) - n\mu K_n(1))E_n] \end{aligned} \qquad (3.120)$$

where $I_n(1)$, $K_n(1)$, $I_n'(1)$ and $K_n'(1)$ represent $I_n(kr_1)$, $K_n(kr_1)$, $I_n'(kr_1)$ and $K_n'(kr_1)$, respectively.

On the other hand, the boundary conditions at $r = r_2$ imply

$$\begin{aligned} I_n(2)D_n + K_n(2)F_n &= -r_2^n \bar{a}_n \\ kr_2 I_n'(2)D_n + kr_2 K_n'(2)F_n &= -n\mu r_2^n \bar{a}_n \end{aligned} \qquad (3.121)$$

$$I_n(2)\,C_n+K_n(2)\,E_n=r_2^{\,n}\bar{b}_n$$
$$kr_2 I_n'(2)\,C_n+kr_2 K_n'(2)\,E_n=n\mu r_2^{\,n}\bar{b}_n \qquad (3.122)$$

which reduce to

$$F_n=-r_2^{\,n}\,(kr_2 I_n'(2)-n\mu I_n(2))\,\bar{a}_n$$
$$D_n=r_2^{\,n}\,(kr_2 K_n'(2)-n\mu K_n(2))\,\bar{a}_n \qquad (3.123)$$
$$E_n=r_2^{\,n}\,(kr_2 I_n'(2)-n\mu I_n(2))\,\bar{b}_n$$
$$C_n=-r_2^{\,n}\,(kr_2 K_n'(2)-n\mu K_n(2))\,\bar{b}_n \qquad (3.124)$$

where 2 in parentheses represents kr_2.

Substituting (3.123) and (3.124) respectively into (3.119) and (3. 120), we obtain

$$2n\mu r_1^{\,n} a_n=r_2^{\,n}\bar{a}_n\Psi_c$$
$$2n\mu r_1^{\,n} b_n=r_2^{\,n}\bar{b}_n\Psi_c \qquad (3.125)$$

where

$$\Psi_c=[kr_1 K_n'(1)+n\mu K_n(1)][kr_2 I_n'(2)-n\mu I_n(2)]$$
$$-[kr_1 I_n'(1)+n\mu I_n(1)][kr_2 K_n'(2)-n\mu K_n(2)]. \qquad (3.126)$$

The ratio of the magnetic potential coefficient for $r_2>r$ to that for $r>r_1$ is then obtained as

$$\bar{a}_n/a_n=\bar{b}_n/b_n=2n\mu q^{-n}\Psi_c^{-1} \qquad (3.127)$$

where

$$q=r_2/r_1. \qquad (3.128)$$

On considering the shielding ratio S_i which has been used in the foregoing sections, it is obvious that

$$S_i=\mathrm{mod}\,(2n\mu q^{-n}\Psi_c^{-1}). \qquad (3.129)$$

We are now in a position to estimate S_i when $\sigma\to 0$ or $k\to 0$. When $z\to 0$, formulas for modified Bessel functions as found in text books of mathematical physics lead to

$$\lim_{z\to 0} I_n(z)\approx\frac{1}{n!}\left(\frac{z}{2}\right)^n$$
$$\lim_{z\to 0} I_n'(z)\approx\frac{1}{2}\frac{1}{(n-1)!}\left(\frac{z}{2}\right)^{n-1}$$
$$\lim_{z\to 0} K_n(z)\approx\frac{(-1)^{n+1}}{n!}\left(\frac{z}{2}\right)^n\log_e\frac{z}{2}+\frac{1}{2}(n-1)!\left(\frac{z}{2}\right)^{-n} \qquad (3.130)$$
$$\lim_{z\to 0} K_n'(z)\approx\frac{(-1)^{n+1}}{(n-1)!}\frac{1}{2}\left(\frac{z}{2}\right)^{n-1}\log_e\frac{z}{2}-\frac{1}{4}n!\left(\frac{z}{2}\right)^{-n-1}.$$

With the aid of (3.130), Ψ_c is calculated as

$$\Psi_c=\frac{(-1)^{n+1}}{[(n-1)!]^2}(\mu^2-1)\left(\frac{z_1}{2}\right)^n\left(\frac{z_2}{2}\right)^n\log_e q$$
$$+\frac{n}{2}[(\mu^2+1)q^{-n}+(\mu-1)^2]. \tag{3.131}$$

The first term of (3.131) vanishes when z_1 and z_2 tend to zero, Ψ_c is then given only by the second term. In that case, we arrive at

$$S_i=\frac{1}{1+\frac{(\mu-1)^2}{4\mu}(1-q^{2n})}. \tag{3.132}$$

If $n=1$, we see that (3.132) agrees with (2.73) which is the shielding ratio for a uniform static field.

3.4.2 Shielding effect to an external field

Putting $n=1$ in (3.129), let us study the shielding effect to a uniform external field. $\mu=1$ is also assumed here.

In that case, (3.125) and (3.126) reduce to

$$\bar{a}_1/a_1=\bar{b}_1/b_1=2q^{-1}\Psi_c^{-1} \tag{3.133}$$

and

$$\Psi_c=k^2r_1r_2[I_0(kr_1)K_2(kr_2)-K_0(kr_1)I_2(kr_2)] \tag{3.134}$$

which are similar to (3.90) and (3.92) for a spherical shell.

With these equations S_i can readily be calculated for various q's by taking Ω as an induction parameter. In a fashion similar to the case of a spherical sheet, we assume a purely periodic variation of angular frequency ω. In that case, we have

$$kr_1=\Omega\sqrt{i}, \qquad kr_2=q\Omega\sqrt{i} \qquad (i=\sqrt{-1})$$

where Ω is given by

$$\Omega=\sqrt{4\pi\sigma\omega}r_1 .$$

The real and imaginary parts of Ψ_c and hence S_i are then calculated for various values of Ω. Tables 3.3 and 3.4 show the real and imaginary parts of S_i for $q=0.8$ and 0.4. The shielding ratio S_i, which is given by mod $\bar{a}_1/a_1$ or mod $\bar{b}_1/b_1$, is calculated along with arg $\bar{a}_1/a_1$ or arg $\bar{b}_1/b_1$ as shown in Figs. 3.9 and 3.10 for $q=0.8$ and 0.4 respectively.

It is interesting to compare Figs. 3.9 and 3.10 to Figs. 3.7 and 3.8

TABLE 3.3 Real and imaginary parts of S_i ($=\bar{a}_1/a_1=\bar{b}_1/b_1$) for q=0.8.

Ω	Re(S_i)	Im(S_i)
0.5	0.9995	−0.0225
0.7	0.9982	−0.0440
1.0	0.9925	−0.0894
1.5	0.9630	−0.1955
2.0	0.8912	−0.3236
2.5	0.7681	−0.4411
3.0	0.6093	−0.5155
4.0	0.3074	−0.5118
5.0	0.1223	−0.4113
7.0	−0.0154	−0.2322
10.0	−0.0523	−0.0974
15.0	−0.0370	−0.0137
20.0	−0.0141	0.0057
30.0	0.0006	0.0024
40.0	0.0005	−0.0001
50.0	0.0000	−0.0001

TABLE 3.4 Real and imaginary parts of S_i ($=\bar{a}_1/a_1=\bar{b}_1/b_1$) for q=0.4.

Ω	Re(S_i)	Im(S_i)
0.5	0.9978	−0.0524
0.7	0.9915	−0.1022
1.0	0.9651	−0.2043
1.5	0.8406	−0.4133
2.0	0.6004	−0.5758
2.5	0.3223	−0.6097
3.0	0.1000	−0.5373
4.0	−0.1144	−0.3082
5.0	−0.1486	−0.1303
7.0	−0.0703	0.0166
10.0	0.0005	0.0162
15.0	0.0011	−0.0009
20.0	−0.0001	0.0000
30.0	0.0000	0.0000
40.0	0.0000	0.0000
50.0	0.0000	0.0000

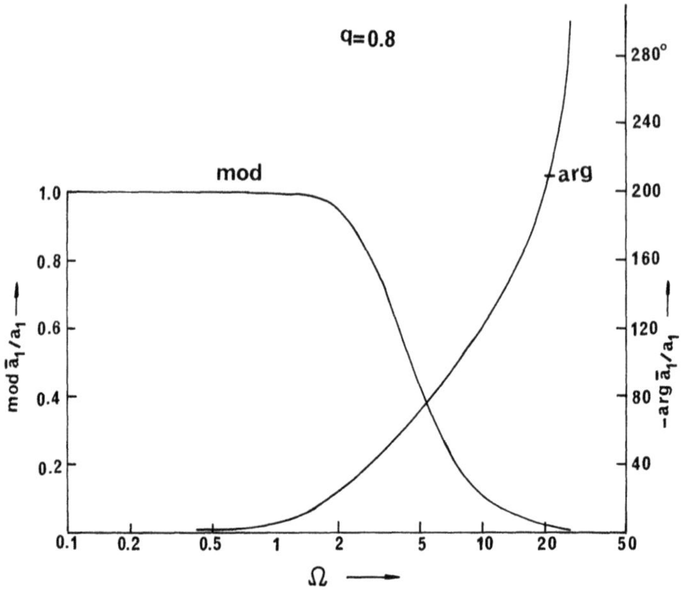

FIG. 3.9 mod $\bar{a}_1/a_1$ and arg $\bar{a}_1/a_1$ as functions of induction parameter Ω for a cylindrical shell of $q=0.8$.

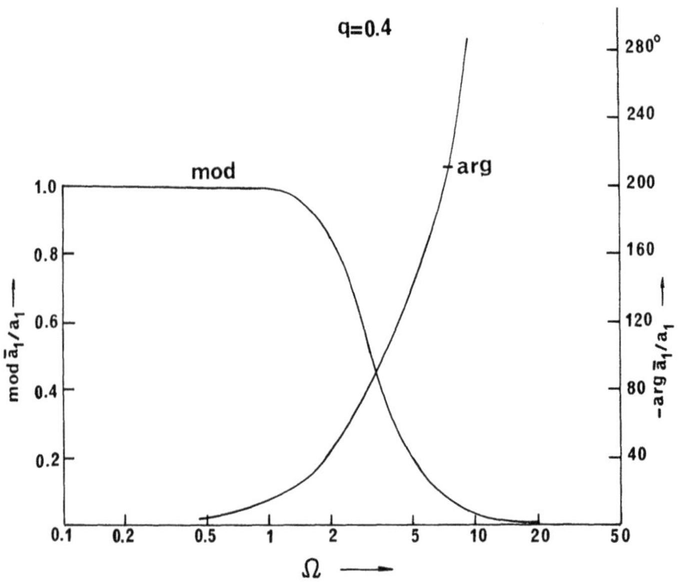

FIG. 3.10 mod $\bar{a}_1/a_1$ and arg $\bar{a}_1/a_1$ as functions of induction parameter Ω for a cylindrical shell of $q=0.4$.

which represent the shielding effect of spherical shells. The ratios of the inner radius to the outer one are taken as 0.8 and 0.4 for both the cylindrical and spherical cases. It is observed that the general features of the change in the shielding ratio as Ω increases are much the same for both cases although the shielding for the cylindrical shell is somewhat larger than that for the spherical one. Since the cylindrical shell is assumed to be infinitely long, it seems natural that the degree of shielding for a cylindrical shell is larger than that for an equivalent spherical shell.

Our discussions concerning applications to shielding by spherical shells may approximately hold good for cylindrical shells, so that an alternating magnetic field of 100 Hz is almost completely shielded by a copper cylindrical shell 10 cm in outer radius and 8 cm in inner radius.

In a fashion similar to Subsection 3.3.4, electromagnetic induction by a magnetic field arising from inside a cylindrical shell can readily be discussed. It is easy to prove that the degree of shielding to an alternating field arising from a line of magnetic dipoles placed at the cylinder axis is exactly the same as that to an external uniform field, although no details are given here.

3.5 Square and Octagonal Shells of Infinite Length

Unlike spherical and cylindrical shells, no analytical approach is possible when studying electromagnetic induction in a conductor of arbitrary shape. In some cases, however, we may use a numerical approach to tackle the problem. For example, the relaxation method can be used for solving problems of electromagnetic induction in a shell having a cross section of rectangular or octagonal shape.

3.5.1 Differential equations and boundary conditions

In order to determine the distribution of magnetic lines of force, let us in this section use magnetic flux function in place of magnetic potential as in a number of two-dimensional problems in Sections 3.2, 3.3 and 3.4. Having been defined by a relation such as (2.25), the magnetic flux function ψ satisfies

$$\nabla^2\psi=0 \tag{3.135}$$

in free space. Meanwhile, we have

$$\nabla^2\psi=4\pi\sigma\mu\frac{\partial\psi}{\partial t} \tag{3.136}$$

in a conductor. (3.136) is equivalent to (3.8) and (3.17).

In order to make time and length dimensionless, we put

$$\begin{aligned} t &= 4\pi\sigma\mu D^2\tau \\ x &= D\xi \\ z &= D\zeta \end{aligned} \tag{3.137}$$

where D is a typical length.

(3.136) is then rewritten as

$$\frac{\partial^2\psi}{\partial\xi^2}+\frac{\partial^2\psi}{\partial\zeta^2}=\frac{\partial\psi}{\partial\tau}. \tag{3.138}$$

When the time-variation is purely periodic with an angular frequency ω, ψ can be separated into real and imaginary parts and written as

$$\psi=(\psi_1+i\psi_2)e^{i\Omega_1{}^2\tau}. \tag{3.139}$$

In that case, we obtain

$$\frac{\partial^2\psi_1}{\partial\xi^2}+\frac{\partial^2\psi_1}{\partial\zeta^2}=-\Omega_1{}^2\psi_2 \tag{3.140}$$

and

$$\frac{\partial^2\psi_2}{\partial\xi^2}+\frac{\partial^2\psi_2}{\partial\zeta^2}=\Omega_1{}^2\psi_1 \tag{3.141}$$

in which

$$\Omega_1{}^2=\mu\Omega^2, \quad \Omega=\sqrt{4\pi\sigma\omega}D \tag{3.142}$$

so that Ω is a kind of induction parameter as defined in the foregoing sections.

Equations (3.140) and (3.141) should hold good in the conductor, while

$$\frac{\partial^2\psi_1}{\partial\xi^2}+\frac{\partial^2\psi_1}{\partial\zeta^2}=0 \tag{3.143}$$

and

$$\frac{\partial^2\psi_2}{\partial\xi^2}+\frac{\partial^2\psi_2}{\partial\zeta^2}=0 \tag{3.144}$$

hold good in free space.

As for the boundary conditions at the interface between the conductor and the free space, the continuity of the tangential magnetic field and the normal magnetic flux density should be satisfied as before. In addition, we have to consider conditions at a boundary where the influence of the induced field practicaly vanishes. Such boundaries are taken at a reasonably large distance from the conductor. Assuming a uniform magnetic field H_0 as the external inducing field, we put

$$\psi_1 = -H_0 z \qquad \psi_2 = 0 \tag{3.145}$$

on the boundary.

3.5.2 *Application of the relaxation method*

Covering the *xz* plane with a square mesh, we can solve Eqs. (3.140), (3.141), (3.143) and (3.144) by use of the relaxation method. It is convenient to assume that D represents the grid interval.

Since (3.140) and (3.141) are coupled with one another, the relaxation procedure becomes a little more complicated than those which appeared in the last chapter. We first assume appropriate values of ψ_2 at each grid point and solve Eq. (3.140) obtaining ψ_1. The values of ψ_1 thus obtained are substituted into the righthand-side of (3.141) solving which ψ_2 can be subsequently obtained. The new set of ψ_2 is then introduced to the righthand-side of (3.140). In such a fashion, the relaxation computation is alternatively and successively applied to (3.140) and (3.141) until the residuals at all the grid points become sufficiently small.

No details of the relaxation procedure will be given here because the actual processes, including the treatment of the boundary conditions, are much the same as those in the last chapter.

3.5.3 *Square prism and shell*

3.5.3.1 *Square prism*

First of all, distributions of the real and imaginary parts of magnetic flux functions for a conducting solid prism of square cross section placed in a uniform alternating magnetic field are obtained by the relaxation method and shown in Fig. 3.11 for various values of Ω. $\mu=1$ is assumed in this case.

No appreciable effect of the conductor can be observed on the distribution of the real part of the magnetic lines of force when Ω is small. It is observed, however, that the magnetic lines of force tend to be deflected from the conductor as Ω increases. When Ω is infinite, no

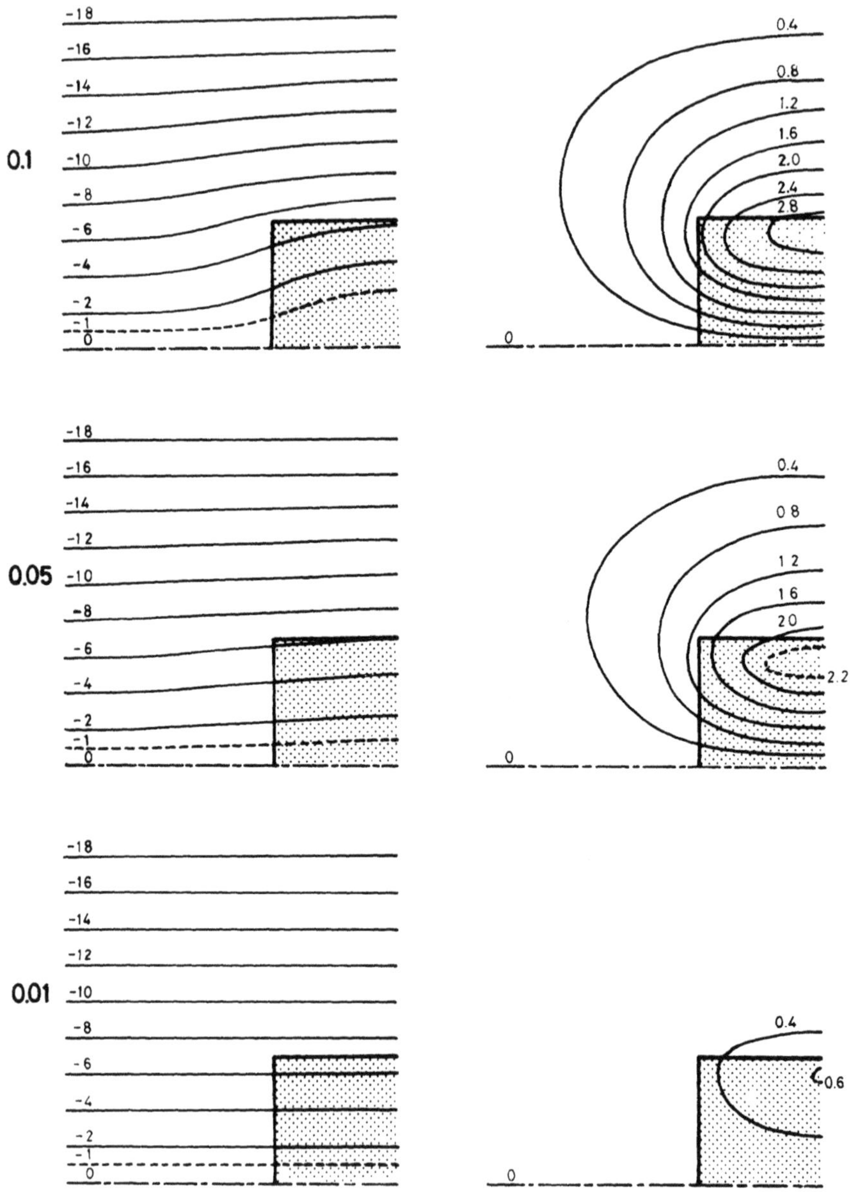

FIG. 3.11(a) Magnetic lines of force around a conducting prism having a square cross section. Taking the induction parameter as a parameter, the real and imaginary parts are shown on the left and right sides, respectively.

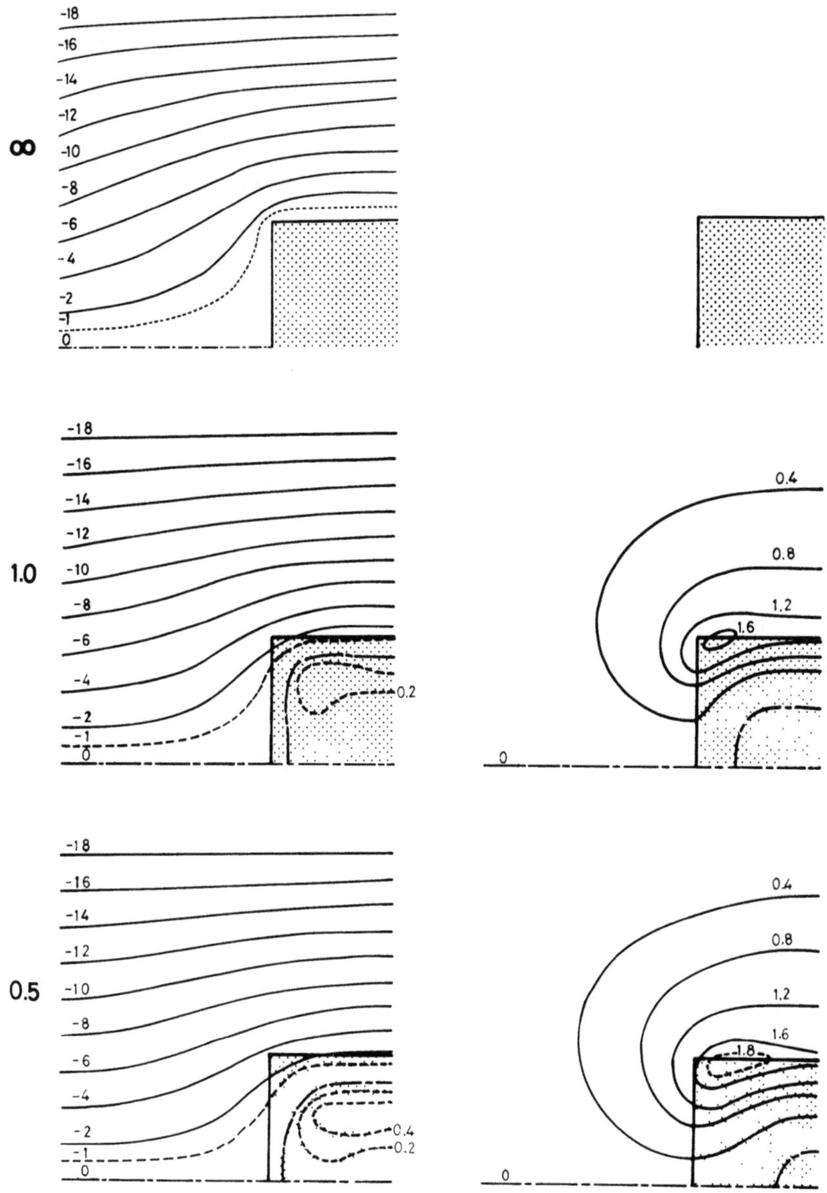

FIG. 3.11(b) Magnetic lines of force around a conducting prism having a square cross section.

magnetic lines of force can enter into the conductor. On the other hand, the imaginary part of the magnetic lines of force tends to grow as Ω increases reaching a maximum somewhere between $\Omega^2=0.05\sim0.1$. For larger values of Ω, the intensity of imaginary part decreases. No imaginary part appears for a perfect conductor. The behaviour of the imaginary part is much the same as the response of a conducting sphere or cylinder placed in a uniform alternating magnetic field (KERTZ, 1960; RIKITAKE, 1966). As can be seen in Fig. 3.11, the imaginary part of the magnetic lines of force is distributed in a similar way to that of a magnetic dipole although the magnetic lines of force tend to encircle the edge of the prism for a high value of induction parameter.

3.5.3.2 Square shell

Next, a square shell as shown in Fig. 3.12 is considered. When $\mu=1$, the distributions of magnetic lines of force are shown in Fig. 3.12 for various values of the induction parameter. In Fig. 3.12, it is clearly demonstrated that the shielding effect becomes highly appreciable for an induction parameter amounting to 1 or so.

Numerical differentiation of the magnetic flux function with respect to z gives the x component of magnetic field. The shielding ratio S_i for the real part of the x component of the magnetic field is estimated as 0.02 at the center for $\Omega^2=1.0$. Although no exact estimate of the imaginary part is made, the modulus of the ratio of the internal field to the external field is of the same order as the value cited above. It is estimated, however, that the modulus amounts to about 0.88 for $\Omega^2=0.1$ resulting in poor shielding.

It should be noticed that the definition of Ω for a cylindrical shell as studied in Subsection 3.4 is different from that of Ω in this subsection. The former is about ten times as large as the latter because $10D$ is equal to the length of one-half of the wall of the shell.

No exact comparison of the shielding power between a square and cylindrical shell having the same thickness is made here. But the mod $\bar{a}_1/a_1$ curve in Fig. 3.9 provides an estimate of S_i for a cylindrical shell which may be compared to that of a square shell of about the same thickness. It is roughly concluded that the shielding power of a square shell is not much different from that of a cylindrical shell of the same thickness as far as the space near the center of the shell is concerned.

Similar investigations into shielding effects were conducted by K. Tsukuda (Unpulished, 1979) for a thicker shell as well as for double square shells. These shells provide a higher rate of shielding than the single shell treated in the above.

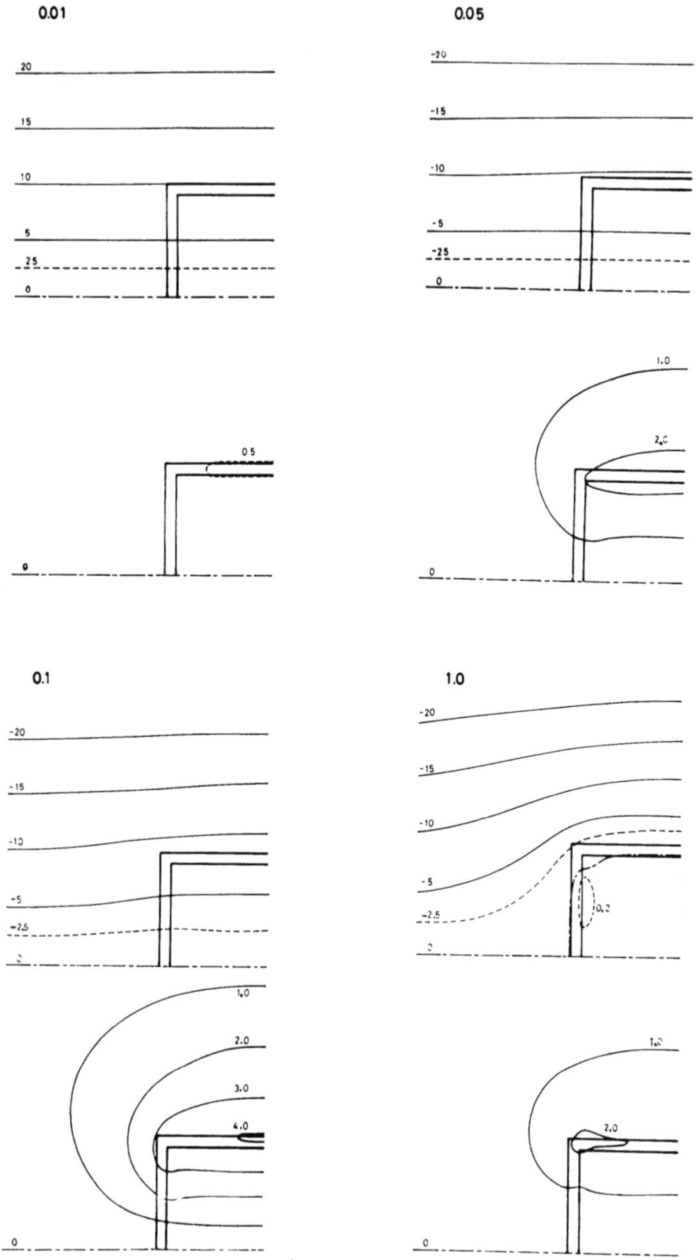

FIG. 3.12 Magnetic lines of force around a conducting square shell for an alternating magnetic field. $\mu=1$ is assumed for the conductor. The induction parameter Ω, which is taken as a parameter, is shown with thick letters for each pair of real and imaginary parts.

It is of theoretical and practical interest to see to what extent an external magnetic field is shielded by a conducting square shell having a high magnetic permeability. Figure 3.13 shows the distribution of magnetic lines of force in and around a square shell, for which the permeability is 100 in electromagnetic units, for various induction parameters.

As can be seen in the figure, the distribution pattern of the real part of the magnetic lines of force is much the same as that for a static case such as shown in Fig. 2.98, although some imaginary part of the magnetic lines of force appears. It is estimated that the shielding ratio for the real part amounts to 0.05 at the center of the shell for $\Omega^2=0.1$, while the imaginary part of the magnetic field there is only about a quarter of the real part.

When $\Omega^2=0.001$, the shielding ratio of the real part of the x component of magnetic field on the z axis changes as the distance from the center increases towards the shell wall, as shown in Fig. 3.14 for $\mu=100$ and 1000. In the figure, z denotes the index which shows the distance from the center. The distance between the outer wall and the center of the shell is divided into ten units. The thickness of the shell equals to 1 unit. It is demonstrated in the figure that the shielding ratio becomes smaller as the shell wall is approached. This was also the case for a static magnetic field as shown in Fig. 2.103. In the present case, the imaginary part is negligibly small.

3.5.3.3 *Square shell in a magnetic field in a direction which makes an angle of 45° with the shell wall*

The distribution of magnetic lines of force is obtained for $\mu=1$ and $\mu=100$ as shown respectively in Figs. 3.15 and 3.16. The characteristics of the distribution of magnetic lines of force thus obtained are much the same as those for the square shell with walls parallel or perpendicular to the external magnetic field. We notice, however, that the leakage of magnetic lines of force at the upper and lower edges is considerable for the ferromagnetic case.

In the case of the non-ferromagnetic shell, the shielding ratio for the real part is estimated as 0.04 at the center for $\Omega^2=1.0$. Assuming $\Omega^2=0.001$, the shielding ratio of the real part of the x component of magnetic field on the z axis is calculated as shown in Fig. 3.17 for $\mu=100$ and 1000. It is remarkable that the ratio increases tremendously at points close to the edge, the magnetic flux density exceeding that outside the shell there. This is quite natural because of the leakage of magnetic lines of force as shown in Fig. 3.16.

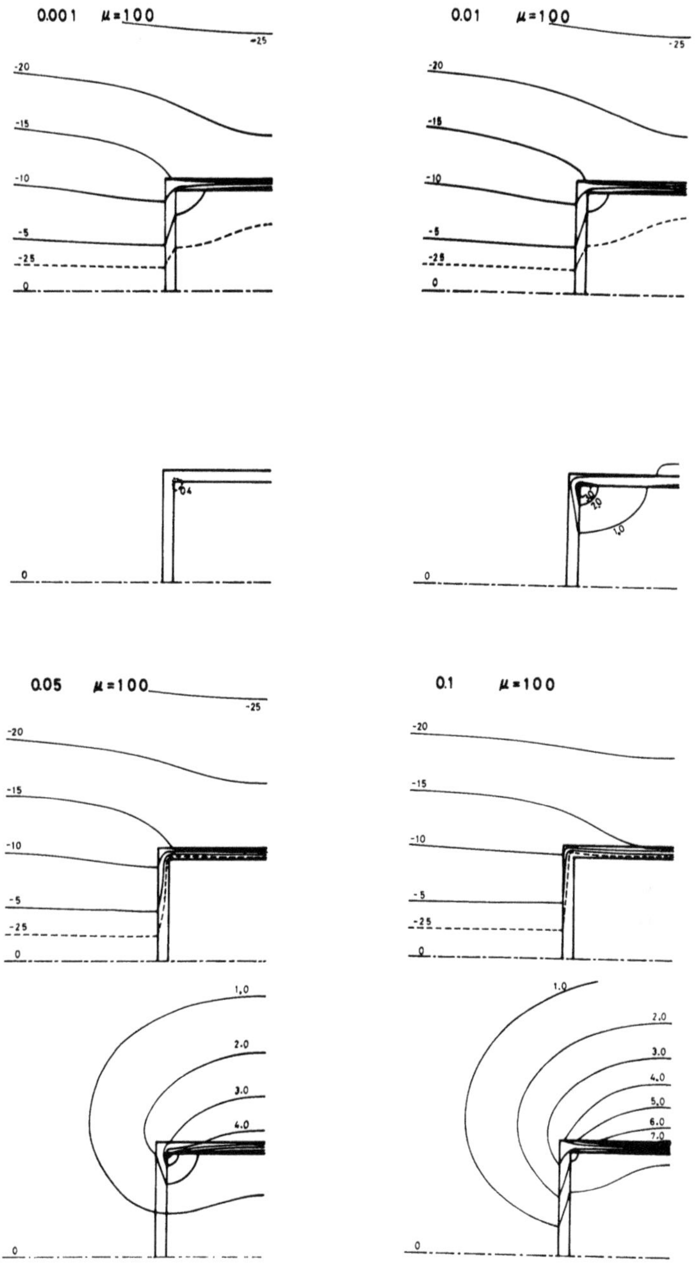

FIG. 3.13 Magnetic lines of force around a conducting as well as magnetically permeable square shell for an alternating magnetic field. $\mu = 100$ is assumed. The induction parameter Ω, which is taken as a parameter, is shown with thick letters for each pair of real and imaginary parts.

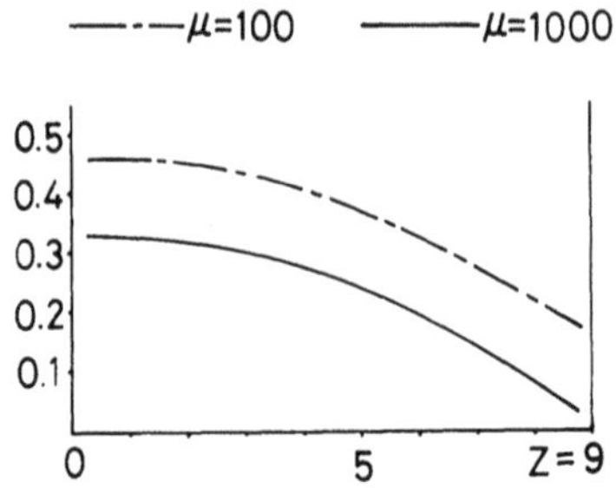

FIG. 3.14 Changes in the shielding ratio of the real part of the x component of magnetic field, which is parallel to the external magnetic field, along the z axis. $\Omega^2=0.001$ is assumed. Two values 100 and 1000 are assumed for μ. The unit of z is given in the text.

3.5.3.4 Octagonal shell

The foregoing studies indicate that leakage of magnetic lines of force around edges and corners of a conducting or ferromagnetic shell plays a vital role in giving rise to local concentrations of magnetic lines of force. It is thus surmised that the cross section of a shell should be as close as possible to a circle in order to have a fairly uniform magnetic field in the space inside the shell.

The behaviour of magnetic lines of force around an octagonal shell as shown in Figs. 3.18 and 3.19 will be studied here by means of the relaxation method. It is supposed that such a model provides a better approximation of a cylindrical shell than does a square shell.

As can be seen in Fig. 3.18, the magnetic lines of force are distributed fairly uniformly inside the shell for the non-ferromagnetic case. Even in the case of ferromagnetic shells, magnetic lines of force are distributed fairly uniformly over a large portion of the space inside the shell. Similarly to Figs. 3.14 and 3.17, the change in the real field intensity of the x component with distance in the z direction from the center is shown in Fig. 3.20 in which we can clearly see that the field component is fairly uniform over a large portion of the space inside the octagonal shell. $\Omega^2=0.001$ in assumed as before. It is understood that the present model is close to a cylindrical shell model, so that the uniformity of the field is roughly achieved inside the shell. It should be said, however, that the degree of shielding in the central region is not much different from that for the square shells studied previously.

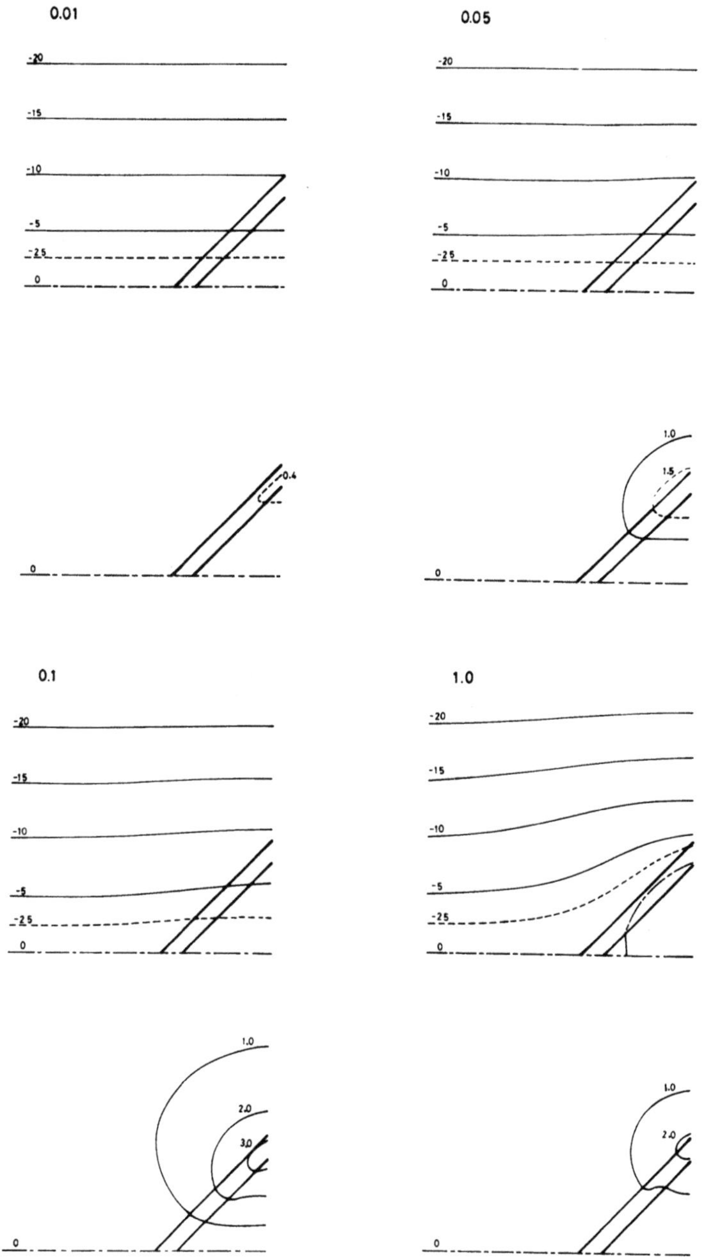

FIG. 3.15 Magnetic lines of force around a conducting square shell ($\mu=1$) in an alternating magnetic field whose direction makes an angle of 45° with the shell wall. Induction parameter Ω is taken as a parameter which is shown with thick letters.

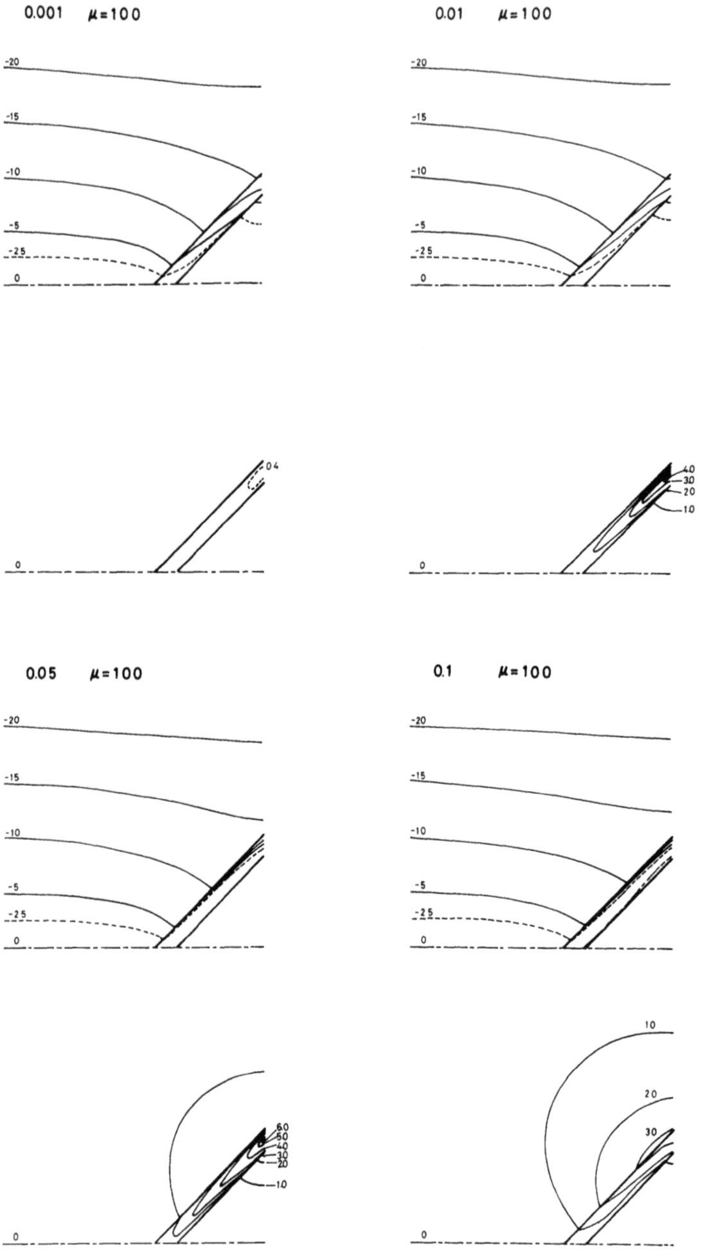

FIG. 3.16 Magnetic lines of force around a conducting square shell ($\mu = 100$) in an alternating magnetic field whose direction makes an angle of 45° with the shell wall. Induction parameter Ω is taken as a parameter which is shown with thick letters.

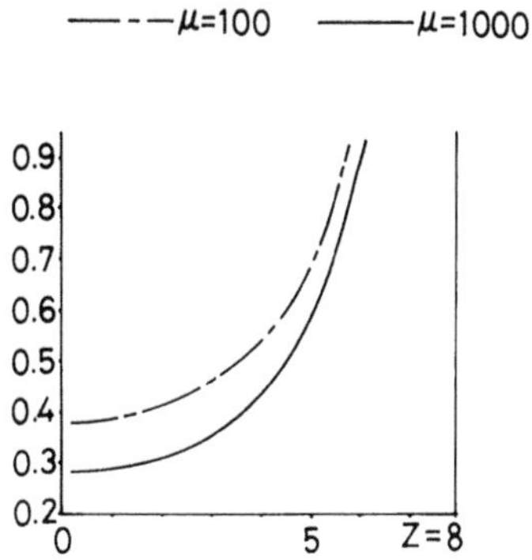

FIG. 3.17 Changes in the shielding ratio of the real part of the x component of magnetic field, which is parallel to the external magnetic field, along the z axis. $\Omega^2=0.001$ is assumed. Two values 100 and 1000 are assumed for μ. The unit of z is given in the text.

3.6 *Thin Sheet or Shell*

There are many problems related to electromagnetic induction in which a conducting sheet or shell can be regarded as one which is infinitesimally thin. Such an approximation becomes possible when the typical wavelength of the inducing magnetic field is much larger than the thickness of the sheet. The theory of electromagnetic induction in a thin sheet can be formulated more easily than that for a general conductor, as developed in Section 3.1, so that the theory can be applied to a sheet in which the electrical conductivity is distributed non-uniformly. Electromagnetic induction in a thin sheet is especially important for a number of geophysical problems such as electromagnetic induction in the ionosphere and in the oceans.

3.6.1 Theory of electromagnetic induction in a thin sheet

The general theory of electromagnetic induction in thin sheets and shells has been developed mostly by PRICE (1949). What follows largely depends on Price's work.

Assuming that displacement currents can be ignored, we write Maxwell's equations as

$$\text{curl}\,\boldsymbol{H}=4\pi\boldsymbol{i} \tag{3.146}$$

$$\text{curl}\,\boldsymbol{E}=-\frac{\partial \boldsymbol{H}}{\partial t} \tag{3.147}$$

$$\boldsymbol{i}=\sigma\boldsymbol{E} \tag{3.148}$$

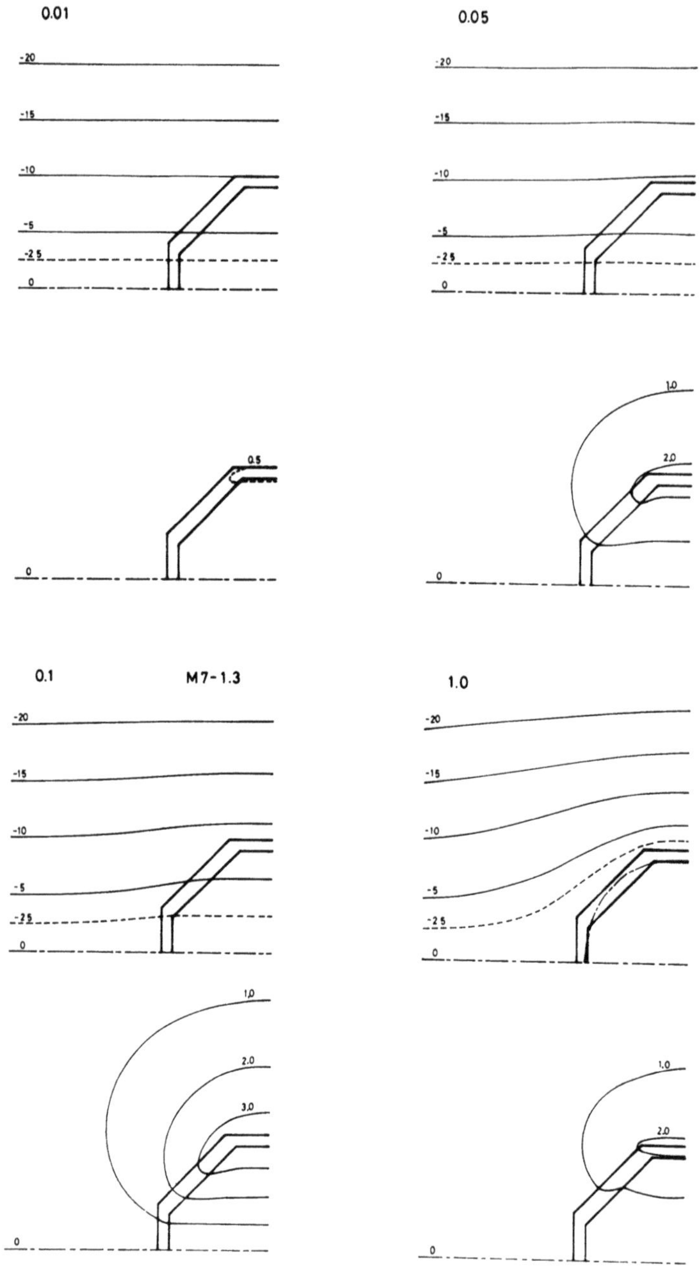

FIG. 3.18 Magnetic lines of force around a conducting octagonal shell ($\mu = 1$) in an alternating magnetic field. Induction parameter Ω is taken as a parameter which is shown with thick letters.

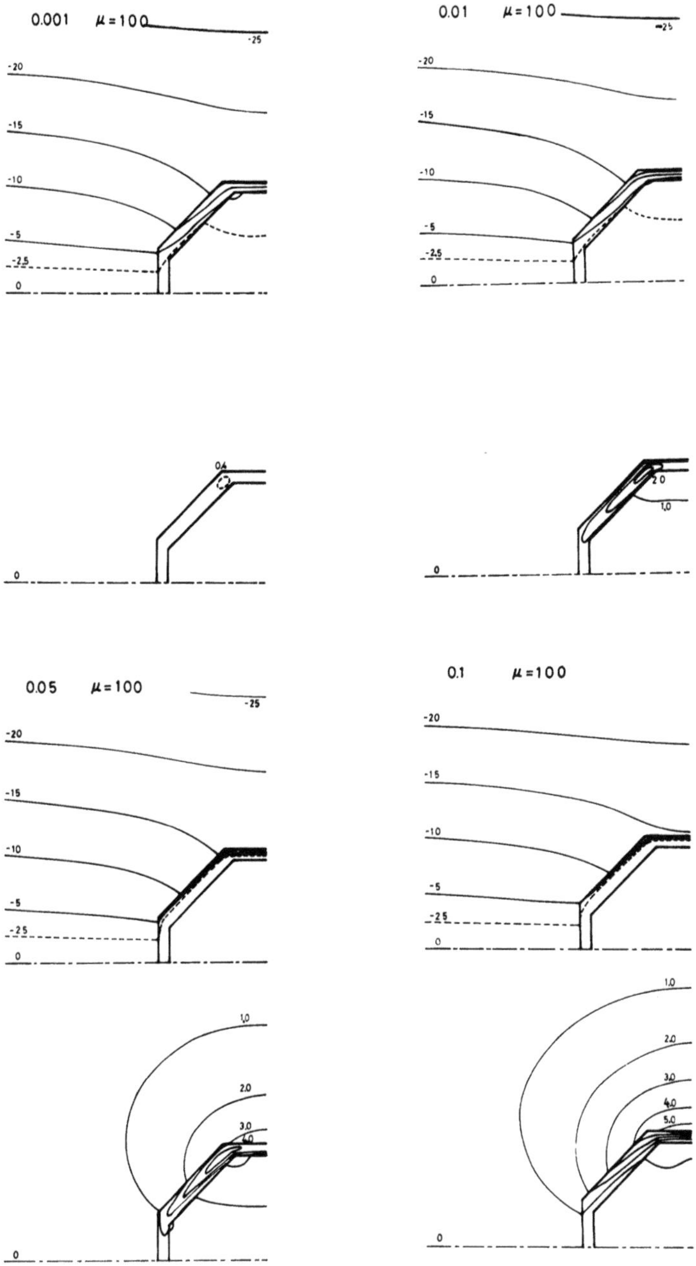

FIG. 3.19 Magnetic lines of force around a conducting octagonal shell (μ= 100) in an alternating magnetic field. Induction parameter Ω is taken as a parameter which is shown with thick letters.

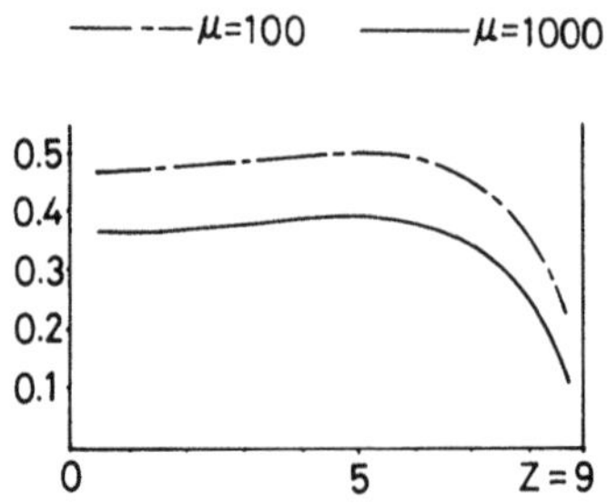

FIG. 3.20 Changes in the shielding ratio of the real part of the x component of magnetic field, which is parallel to the extertal magnetic field, along the z axis. $\Omega^2=0.001$ is assumed. Two values 100 and 1000 are assumed for μ. The unit of z is given in the text.

in which the electrical conductivity σ is in general a function of the space variables. The notations used here are conventional and are those adopted in Section 3.1. The magnetic permeability is assumed unity throughout this section.

Integrating (3.148) over the thickness L of the sheet, we define the integrated electric current as

$$\boldsymbol{i}_s=\int_0^L \boldsymbol{i}\,d\zeta.$$

Similarly, the integrated conductivity, which can be expressed by the reciprocal of electric resistance ρ, can be defined by

$$\frac{1}{\rho}=\int_0^L \sigma\,d\zeta.$$

If we denote the component of $\boldsymbol{E}$ parallel to the sheet by $\boldsymbol{E}_s$, we obtain

$$\boldsymbol{E}_s=\rho\boldsymbol{i}_s\,. \tag{3.149}$$

Applying (3.147) to a small circuit in the surface of the sheet, we obtain

$$\text{curl}\,\boldsymbol{E}_s=-\boldsymbol{n}\,\frac{\partial H_n}{\partial t} \tag{3.150}$$

where $\boldsymbol{n}$ is a unit vector normal to the surface from the negative to the positive side. If $\boldsymbol{E}_s$ is eliminated from (3.150) by virtue of (3.149), we obtain

$$\rho \text{ curl } \boldsymbol{i}_s + \text{grad } \rho \wedge \boldsymbol{i}_s = -\boldsymbol{n} \frac{\partial H_n}{\partial t}. \tag{3.151}$$

Next, we apply the circuital relation (3.146) to a small rectangular circuit which has two sides of length δ_s parallel to, and on either side of, the sheet. We then obtain relation

$$(H_s)_+ \delta_s - (H_s)_- \delta_s = 4\pi i_s \delta_s \sin\theta$$

where the + and − suffixes indicate values at the positive and negative faces of the sheet. θ denotes the angle between δ_s and $\boldsymbol{i}_s$. We, therefore, obtain

$$\boldsymbol{i}_s = (1/4\pi)\boldsymbol{n} \wedge (\boldsymbol{H}_+ - \boldsymbol{H}_-). \tag{3.152}$$

It is sometimes convenient to make use of a stream-line function for the currents flowing in the sheet. Ψ, which is often called the current function, is defined as

$$\Psi = (1/4\pi)(\boldsymbol{W}_+ - \boldsymbol{W}_-) \tag{3.153}$$

where $\boldsymbol{W}$ is the magnetic potential defined in the free space. (3.152) is then rewritten as

$$\boldsymbol{i}_s = -\boldsymbol{n} \wedge \text{ grad } \Psi. \tag{3.154}$$

Introducing $\boldsymbol{i}_s$ of (3.152) into (3.151), we have

$$\rho \text{ curl } [\boldsymbol{n} \wedge (\boldsymbol{H}_+ - \boldsymbol{H}_-)] + \text{grad } \rho \wedge [\boldsymbol{n} \wedge (\boldsymbol{H}_+ - \boldsymbol{H}_-)] = -4\pi \boldsymbol{n} \frac{\partial H_n}{\partial t} \tag{3.155}$$

which reduces to

$$\rho \text{ div } (\boldsymbol{H}_+ - \boldsymbol{H}_-) + \text{grad } \rho \cdot (\boldsymbol{H}_+ - \boldsymbol{H}_-) = -4\pi \frac{\partial H_n}{\partial t}. \tag{3.156}$$

When there is a varying magnetic field arising from external sources, it is convenient to denote the potential due to this source by W_e and that due to the induced currents in the sheet by W_i, so that $W = W_e + W_i$.

The inducing field or grad W_e is continuous across the sheet, so that (3.156) becomes

$$\begin{aligned} \rho \text{ div grad } (\boldsymbol{W}_{i+} - \boldsymbol{W}_{i-}) + \text{grad } \rho \cdot \text{ grad } (\boldsymbol{W}_{i+} - \boldsymbol{W}_{i-}) \\ = -4\pi \frac{\partial}{\partial t}\left(\frac{\partial W_e}{\partial n} + \frac{\partial W_i}{\partial n}\right). \end{aligned} \tag{3.157}$$

The problem of finding the field produced by electric currents induced in a thin sheet by a given inducing potential W_e is thus reduced to the problem of solving

$$\nabla^2 W_i = 0 \tag{3.158}$$

outside the sheet on the condition that W_i vanishes at infinity and satisfies (3.157) at the sheet.

3.6.2 Plane sheet

Let us assume that a plane sheet is placed at $z=0$ in a rectangular coordinate system, it is obvious that

$$W_{i+} = -W_{i-} \tag{3.159}$$

and hence

$$\Psi = (1/2\pi) W_{i+} \tag{3.160}$$

so that (3.157) reduces to

$$-\rho\left(\frac{\partial^2 W_i}{\partial z^2}\right)_+ + \text{grad}\,\rho \cdot \text{grad}\,\mathbf{W}_{i+} = -2\pi\frac{\partial}{\partial t}\left(\frac{\partial W_e}{\partial z} + \frac{\partial W_i}{\partial z}\right)_+. \tag{3.161}$$

For a uniform sheet, (3.161) simplifies to

$$\rho\left(\frac{\partial^2 W_i}{\partial z^2}\right)_+ = 2\pi p\left(\frac{\partial W_e}{\partial z} + \frac{\partial W_i}{\partial z}\right)_+ \tag{3.162}$$

where $\partial/\partial t$ is replaced by operator p.

Supposing that an inducing field arises from sources lying in a space $z>0$, a typical potential can be written as

$$W_e = ee^{kz}\sin kx. \tag{3.163}$$

On the other hand, the induced potential may be written as

$$W_i = \begin{cases} ie^{-kz}\sin kx & (z>0) \\ -ie^{kz}\sin kx & (z<0). \end{cases} \tag{3.164}$$

The boundary condition at $z=0$ then implies

$$i = \frac{2\pi p/k}{\rho + 2\pi p/k}\,e. \tag{3.165}$$

The total potential $W\,(=W_e+W_i)$ for $z<0$ can be obtained as

$$W = \bar{e}e^{kz}\sin kx \tag{3.166}$$

where

$$\bar{e} = \frac{\rho}{\rho + 2\pi p/k} e. \tag{3.167}$$

When the variation is purely periodic with angular frequency ω, we have

$$\frac{\bar{e}}{e} = \frac{1}{1 + 2\pi i \omega \sigma L/k} \tag{3.168}$$

in which a relation $1/\rho = \sigma L$ is assumed, where the conductivity and thickness of the sheet are given by σ and L, respectively.

With the aid of (3.39), (3.168) can be rewritten as

$$\frac{\bar{e}}{e} = \frac{1}{1 + i\Omega^2/2kL}. \tag{3.169}$$

Similarly to (3.36) and (3.37), mod $\bar{e}/e$ and arg $\bar{e}/e$ give the amplitude ratio, or shielding ratio, and the phase difference of the magnetic field between the positive and negative sides of the sheet.

Mod $\bar{e}/e$ and arg $\bar{e}/e$ change with Ω as shown in Fig. 3.21 when $k=0.1$. It is remarkable that both the curves in the figure agree very well with the curves in Fig. 3.2 which were obtained for a plane slab having a finite thickness. We, therefore, see that a thin sheet is a good approximation when studying an electromagnetic induction problem in a plane sheet as long as the thickness of the sheet amounts to

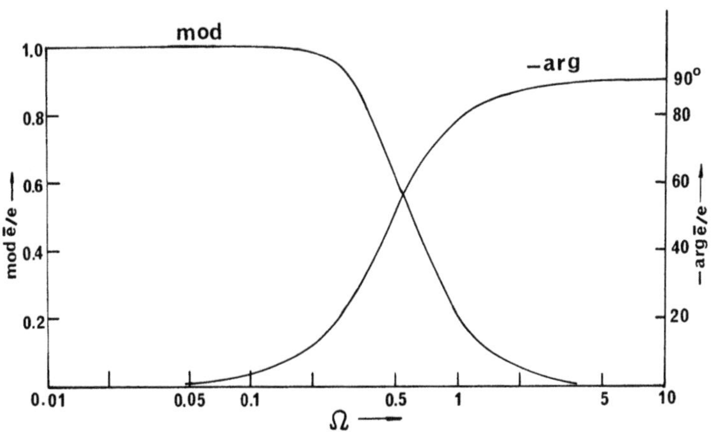

FIG. 3.21 Changes in mod $\bar{e}/e$ and arg $\bar{e}/e$ with Ω for a thin plane sheet, when $k=0.1$.

one-fifth or smaller of the wavelength of the inducing field. Under these circumstances, what we concluded in Section 3.2 about the shielding effect of a plane slab seems to hold good even for the present thin sheet.

3.6.3 Cylindrical sheet

Let us consider a thin cylindrical sheet having a circular cross section. The z axis of cylindrical coordinates (r, θ, z) coincides with the cylindrical axis. In that case, Equation (3.157) can be written as

$$\rho\nabla^2 (W_{i+}-W_{i-})=-4\pi \frac{\partial^2}{\partial t \partial r} (W_e+W_i) \qquad (3.170)$$

where $1/\rho$ is the integrated conductivity as before. ρ is constant everywhere.

Writing (3.170) in cylindrical coordinates and assuming that the radius of the sheet is a, we have

$$\rho \frac{\partial^2}{\partial \theta^2} (W_{i+}-W_{i-})=-4\pi a^2 \frac{\partial^2}{\partial t \partial r} (W_e+W_i) \qquad (3.171)$$

which can be rewritten as

$$\rho \frac{\partial^2 \Psi}{\partial \theta^2}=a^2 \frac{\partial H_r}{\partial t}. \qquad (3.172)$$

We write typical magnetic potentials for $r>a$ and $r<a$, respectively, as W_n and W_n'. They are written as

$$W_n=a_n r^n \cos n\theta+A_n r^{-n} \cos n\theta \qquad (r>a) \qquad (3.173)$$

and

$$W_n'=\bar{a}_n r^n \cos n\theta \qquad (r<a). \qquad (3.174)$$

In (3.173) and (3.174), only terms proportional to $\cos n\theta$ are taken. But we can develop similar relations for the terms expressed by $\sin n\theta$.

Corresponding to (3.173) and (3.174), the current function can be written as

$$\Psi_n=K_n \cos n\theta. \qquad (3.175)$$

The condition for a potential jump between the plus and minus sides of the sheet as given by (3.153) implies

$$a^n a_n+a^{-n}A_n-a^n\bar{a}_n=4\pi K_n. \qquad (3.176)$$

Meanwhile, the continuity condition of the normal magnetic field yields

$$a^{n-1}a_n - a^{-n-1}A_n = a^{n-1}\bar{a}_n. \tag{3.177}$$

In addition to the above two conditions, (3.172) leads to

$$\rho n K_n = a^{n+1}\bar{a}_n p \tag{3.178}$$

where p is a time operator replacing $\partial/\partial t$ as usual.

Eliminating A_n and K_n from the above three equations, we obtain

$$\frac{\bar{a}_n}{a_n} = \frac{n\rho}{n\rho + 2\pi a p}. \tag{3.179}$$

When the time variation is purely periodic with angular frequency ω, we put

$$p = i\omega \qquad (i = \sqrt{-1})$$

as before. In that case, (3.179) becomes

$$\frac{\bar{a}_n}{a_n} = \frac{n}{n + i\Omega^2(L/2a)} \tag{3.180}$$

in a manner similar to the last subsection, where L is the thickness of the sheet. Ω is given by $\Omega = \sqrt{4\pi\sigma\omega a}$.

Assuming $n=1$ and $L/2a=0.1$, $\bar{a}_1/a_1$, which is the shielding ratio of the cylindrical sheet for a uniform field, is calculated and shown in Fig. 3.22 for mod $\bar{a}_1/a_1$ and arg $\bar{a}_1/a_1$.

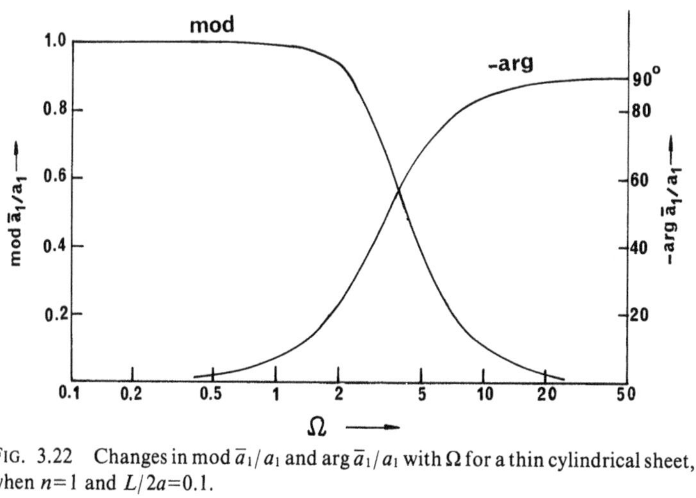

FIG. 3.22 Changes in mod $\bar{a}_1/a_1$ and arg $\bar{a}_1/a_1$ with Ω for a thin cylindrical sheet, when $n=1$ and $L/2a=0.1$.

We see that (3.180) is very similar to (3.169), which gives the shielding ratio for a plane sheet, although the parameters involved are different from one another. The similarity between the shielding ratio versus induction parameter curves in Figs. 3.21 and 3.22 can thus be understood.

We are now in a position to compare the shielding ratio of a thin cylindrical sheet to that of a cylindrical shell having a finite thickness which studied in Section 3.4.

The integrated conductivity of the shell for which the modulus and argument of the shielding ratio are shown in Fig. 3.9 is the same as that of the thin shell for which the same quantities are illustrated in Fig. 3.22. In other words, the curves in Fig. 3.22 represent the results obtained using the thin sheet approximation, the exact solutions being given by the curves shown in Fig. 3.9. It is interesting to note that the absolute values or modulus of the shielding ratio is almost the same for the two treatments although the thin sheet approximation gives a slightly higher shielding rate. It is observed, however, that the phases are very different for the two models when the induction parameter is large. For an infinitesimally thin shell, the phase difference or argument tends to $\pi/2$ when $\Omega\to\infty$. But this is not the case for a shell of finite thickness because the phases are different at different radii in the shell. However, this is of no practical importance because such a discrepancy becomes appreciable only for Ω equal to 10 or larger, when the field inside the shell becomes very small owing to electromagnetic shielding.

3.6.4 Spherical sheet

In the case of induction in a spherical sheet of radius a, the general equation at the sheet given by (3.157) becomes

$$a^{-2}\left\{\frac{\partial\rho}{\partial\theta}\frac{\partial}{\partial\theta}+\frac{1}{\sin^2\theta}\frac{\partial\rho}{\partial\phi}\frac{\partial}{\partial\phi}+\frac{\rho}{\sin\theta}\frac{\partial}{\partial\theta}\left(\sin\theta\frac{\partial}{\partial\theta}\right)+\frac{\rho}{\sin^2\theta}\frac{\partial^2}{\partial\phi^2}\right\}$$
$$\cdot(W_{i+}-W_{i-})=-4\pi\frac{\partial^2}{\partial t\partial R}(W_e+W_i) \qquad (3.181)$$

where the origin of a polar spherical coordinate system (R, θ, ϕ) is taken at the center of the sphere.

The condition for a uniform spherical sheet is obtained from (3.181) by putting $\partial\rho/\partial\theta=\partial\rho/\partial\phi=0$. The condition is then written as

$$\left[\frac{1}{\sin\theta}\frac{\partial}{\partial\theta}\left(\sin\theta\frac{\partial}{\partial\theta}\right)+\frac{1}{\sin^2\theta}\frac{\partial^2}{\partial\phi^2}\right]\Psi=a^2K\frac{\partial H_R}{\partial t} \qquad (R=a) \qquad (3.182)$$

where the integrated conductivity of the sheet is written as K in place of $1/\rho$. Ψ is the current function, as defined in (3.153), and is related to the components of electric current in the sheet by

$$i_{s,\theta}=\frac{\partial\Psi}{a\sin\theta\partial\phi}, \qquad i_{s,\phi}=-\frac{\partial\Psi}{a\,\partial\theta}. \tag{3.183}$$

Let us denote magnetic potentials outside and inside the sheet by W and W' respectively. Their typical terms are given as

$$W_n^m=a\left\{e_n^m\left(\frac{R}{a}\right)^n+i_n^m\left(\frac{R}{a}\right)^{-n-1}\right\}S_n^m \qquad (R>a)$$

and (3.184)

$$W_n^{m'}=a\bar{e}_n^m\left(\frac{R}{a}\right)^n S_n^m \qquad (R<a)$$

where S_n^m stands for the surface hermonic $P_n^{\mathrm{m}}(\cos\theta)\ {}^{\cos}_{\sin}m\phi$. If we express a typicla term of the current function as

$$\Psi_n^m=K_n^m S_n^m \tag{3.185}$$

the jump in the magnetic potential due to the current sheet implies

$$e_n^m+i_n^m-\bar{e}_n^m=\frac{4\pi}{a}K_n^m. \tag{3.186}$$

The continuity condition of the normal magnetic field leads to

$$ne_n^m-(n+1)i_n^m=n\bar{e}_n^m. \tag{3.187}$$

In addition to the above two equations, (3.182) implies

$$n(n+1)K_n^m=a^2Kp\,[ne_n^m-(n+1)i_n^m] \tag{3.188}$$

where p replaces $\partial/\partial t$ as usual.

Eliminating i_n^m and K_n^m from the above three equations, we obtain

$$\frac{\bar{e}_n^m}{e_n^m}=\frac{2n+1}{2n+1+4\pi apK}. \tag{3.189}$$

Assuming a periodic vatiation with an angular frequency ω as before, (3.189) can be rewritten as

$$\frac{\bar{e}_n^m}{e_n^m}=\frac{2n+1}{2n+1+i\Omega^2(L/a)} \tag{3.190}$$

where L is the thickness of the sheet and the induction parameter Ω is given by $\Omega=\sqrt{4\pi\sigma\omega a}$.

When we assume a uniform field, (3.190) reduces to

$$\frac{\bar{e}_1^{\,0}}{e_1^{\,0}} = \frac{1}{1+i\Omega^2(L/3a)} \tag{3.191}$$

by putting $n=1$ and $m=0$.

Assuming $L/a=0.2$, mod $\bar{e}_1^{\,0}/e_1^{\,0}$ and arg $\bar{e}_1^{\,0}/e_1^{\,0}$ are calculated for various values of the induction parameter Ω and are shown in Fig. 3.23. The shielding ratio which is given by mod $\bar{e}_1^{\,0}/e_1^{\,0}$ takes on a value approximately equal to 1 when Ω is smaller than 1. It changes steeply for values of Ω between 2 and 10. Corresponding to the rapid change in the shielding ratio with Ω, the phase difference or $-\arg \bar{e}_1^{\,0}/e_1^{\,0}$ changes from appeoximately 0° to 90°.

It is interesting to compare Fig. 3.23 to Fig. 3.7 for a spherical shell of finite thickness. The integrated conductivity is the same for both models. The comparison indicates that the shielding ratios for the two models approximately agree with each other, to within 10%. We, therefore, see that a thin spherical sheet model provides a good approximation to a spherical sheet model of finite thickness as far as the shielding effect on the amplitude of an alternating magnetic field is concerned. It should be pointed out, however, that the thin sheet model gives a slightly higher shielding rate than the model of finite thickness. We should also note that the phase differences disagree for large values

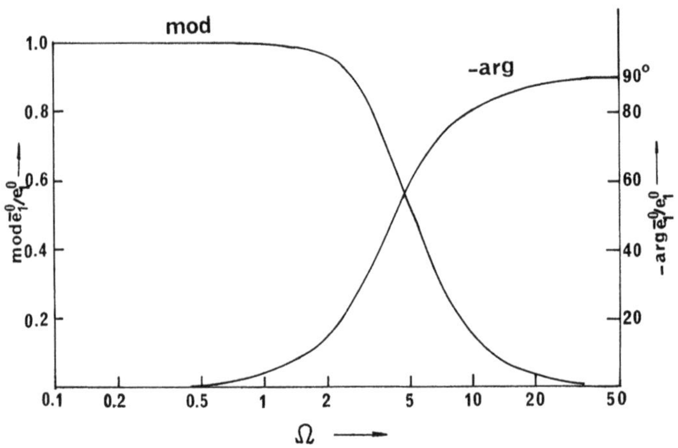

FIG. 3.23 Changes in mod $\bar{e}_1^{\,0}/e_1^{\,0}$ and arg $\bar{e}_1^{\,0}/e_1^{\,0}$ with Ω for a thin spherical sheet. $n=1$, $m=0$ and $L/a=0.2$ are assumed.

of Ω although the amplitude of the magnetic field inside the space surrounded by the sheet is very small for such Ω's.

The above discussion indicates that what we estimated about the electromagnetic shielding power of a copper spherical shell as given in Subsection 3.3.3 may well hold good even for an equivalent thin sheet.

Let us next think of a worldwide ocean covering the earth's surface to a depth of 1000 m. Taking $\sigma=4$ S/m and $a=6400$ km, a uniform external magnetic field having a period of 1-day, 1-hour or 1-minute is shielded by the sheet, the amplitude of the shielded field amounting to 0.790, 0.053 and 0.000 of the original one, respectively. We therefore see that a 1-minute variation is shielded by the hypothetical ocean almost completely.

3.6.5 Hemispherical sheet

As mentioned in Subsection 3.6.1, the general theory of electromagnetic induction in thin shells and sheets has been developed by PRICE (1949). A brief account of the theory is reproduced in Subsection 3.6.1. ASHOUR and PRICE (1948) studied electromagnetic induction in a non-uniform ionosphere. Since then, various approaches to electromagnetic induction in a non-uniform thin sheet, either plane or spherical, have been made mostly in relation to induction in an ocean (e.g., RIKITAKE, 1967, 1968; BULLARD and PARKER, 1970; HOBBS and PRICE, 1970; HOBBS, 1971, 1975; HOBBS and BRIGNALL, 1976; HUTSON *et al.*, 1972, 1973; HEWSON-BROWNE *et al.*, 1973; KENDALL, 1978). These contributions are mainly concerned with the distribution of electric currents induced in an ocean and the resulting magnetic fields. The results play an important role in interpreting geomagnetic changes that are likely to be affected by induced currents in an actual ocean. An excellent review on the topic has recently been published by FAINBERG (1980). Though interesting, the above investigations have not emphasized the shielding effect in the sense studied in this book. In Chapter 4, the resonance-like response of plane and spherical non-uniformly conducting sheets will be presented.

RIKITAKE and YOKOYAMA (1955) were the first to tackle electromagnetic induction in a hemispherical thin sheet by a uniform field. As the induced electric currents and magnetic fields are approximated by a sum of spherical harmonic functions of finite number in this study, no accurate results are obtained at the edge of the sheet where the current density becomes infinitely large.

ASHOUR (1965a, b) succeeded in obtaining an exact solution for a problem of electromagnetic induction in a perfectly-conducting hemispherical shell by a uniform magnetic field, by making use of a suitable transformation of Bessel functions appearing in the potential theory. Taking the radius of the sheet as a, we assume that the sheet occupies the portion $0 \leqq \theta \leqq \pi/2$ and that the inducing field H_0 is directed parallel to the $\theta = 0$ axis. Ashour showed that the current function Ψ is given by

$$\Psi = \begin{cases} (-H_0 a/4\pi^2)\{(3\cos\theta + 1)\tan^{-1}(\cos\theta)^{1/2} + 3(\cos\theta)^{1/2}\} & 0 \leq \theta \leq \pi/2 \\ 0 & \pi/2 < \theta \leq \pi . \end{cases} \qquad (3.192)$$

The analytical expressions of electric currents and tangential and normal magnetic field components, either induced or total, on the sheet are also given by Ashour although they are not shown here. Ashour also presented the distribution of magnetic lines of force as reproduced in Fig. 3.24, in which we see that the magnetic lines of force tend to be deflected from the conductor.

When the inducing field is perpendicular to the $\theta = 0$ axis, the current function becomes

$$\Psi = \begin{cases} (-H_0 a/4\pi^2)[3\sin\theta\,\{\tan^{-1}(\cos\theta)^{1/2} + (\cos\theta)^{1/2}/(1+\cos\theta)\} \\ \quad +2(\pi+2)^{-1}\{2\cot(\theta/2)\tan^{-1}(\cos\theta)^{1/2} \\ -\pi\operatorname{cosec}\theta\,(\cos\theta)^{1/2}\}]\cos\phi & 0 \leq \theta \leq \pi/2 \\ 0 & \pi/2 < \theta \leq \pi \end{cases} \qquad (3.193)$$

and the magnetic lines of force are distributed as shown in Fig. 3.25. In the figure, we can see very clearly some rarefaction of magnetic lines of force although no exact values of shielding by the hemispherical sheet have been given. The extent of the shielding can roughly be estimated from the magnetic flux density in Figs. 3.24 and 3.25. It should be pointed out that the magnetic field at the very edge becomes infinitely large in this case.

DOSS and ASHOUR (1971) extended the above theory to the case of finite conductivity. An iterative procedure starting from the solution for perfect conductivity as originally proposed by PRICE (1949) was adopted.

3.6.6 *Square sheet*

Electromagnetic induction in a thin sheet of rectangular or square

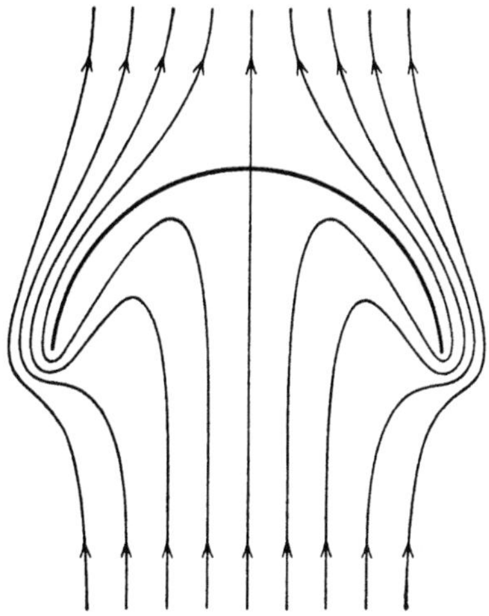

FIG. 3.24 Magnetic lines of force around a perfectly-conducting hemispherical sheet. The field is parallel to the $\theta=0$ axis at a distance (ASHOUR, 1965a).

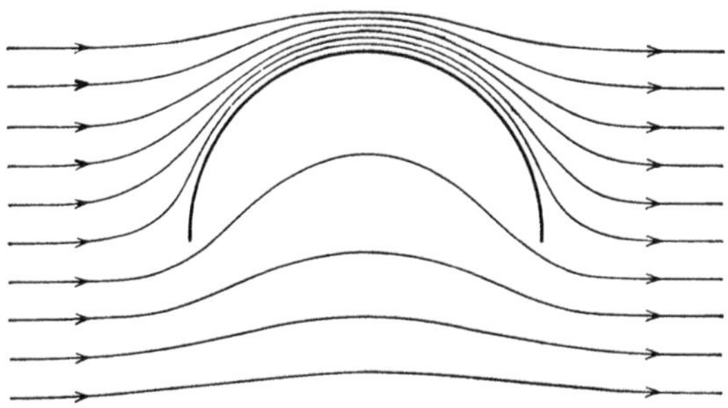

FIG. 3.25 Magnetic lines of force around a perfectly-conducting hemispherical sheet. The field is parallel to the $\theta=\pi/2$ and $\phi=0$ plane at a distance (ASHOUR, 1965a).

shape can be studied by applying the relaxation method to the problem in a manner similar to Section 3.5. The thickness L of the sheet is much smaller than a typical wavelength of the inducing field. The relaxation procedure for the magnetic flux function ψ is more or less the same as that studied in Section 3.5 except the boundary condition.

Let us go back to (3.136) which can be rewritten as

$$\left(\frac{\partial^2}{\partial x^2}+\frac{\partial^2}{\partial z^2}\right)\psi=4\pi\sigma\mu\,\frac{\partial\psi}{\partial t}. \tag{3.194}$$

We further assume that the sheet lies in a plane defined by $z=0$. Integrating both the sides of Eq. (3.19) with respect to z from $z=0$ to $z=L$, we obtain

$$\left(\frac{\partial \psi}{\partial z}\right)_+ - \left(\frac{\partial \psi}{\partial z}\right)_- = 4\pi K\mu \frac{\partial \psi}{\partial t} \tag{3.195}$$

where K is the integrated conductivity of the sheet (COX and FILLOUX, 1974). This is the boundary condition to be adopted at the sheet.

The distribution of magnetic lines of force around a square sheet, which approximates the square shell shown in Fig. 3.12, is illustrated in Fig. 3.26. In the figure, we observe that the distribution of magnetic lines of force for a thin sheet is very similar to that for a square shell as shown in Fig. 3.12. $\mu=1$ is assumed in the computation.

We may therefore conclude that a thin sheet model provides a good approximation to a shell of finite thickness, when the lateral dimensions of the shell are considerably larger than the wall thickness, as far as electromagnetic induction is concerned. The conclusions in Section 3.5 concerning the shielding power of a conducting square shell approximately hold good even for a thin square shell.

3.7 *Shielding by Two Concentric Spherical Shells, One Electrically Conducting, the Other Magnetically Permeable*

As we have seen in this and last chapters, the shielding of an alternating magnetic field by an electrically conducting shell or sheet is intrinsically different from that of a static magnetic field by a magnetically permeable shell or sheet. In the case of a conductor, the magnetic lines of force tend to be deflected from the conductor, so that the magnetic flux density inside the shell becomes smaller. On the other hand, a ferromagnetic shell of high permeability attracts magnetic lines of force in the space surrounding it. As the magnetic lines of force run mostly through the shell, and leak very little outside it, the magnetic flux density in the space surrounded by the shell becomes smaller.

It is clear that the above two effects should be applied simultaneously for efficient shielding to an alternating magnetic field of low frequency. A typical application is certainly the shielding of electronic instruments from 50–60 Hz noise fields arising from commercial electric power lines. In practice, the space to be shielded is first surrounded by a thin highly-permeable cylindrical shell of permalloy. Such a system is then surrounded by a shell of conducting metal of

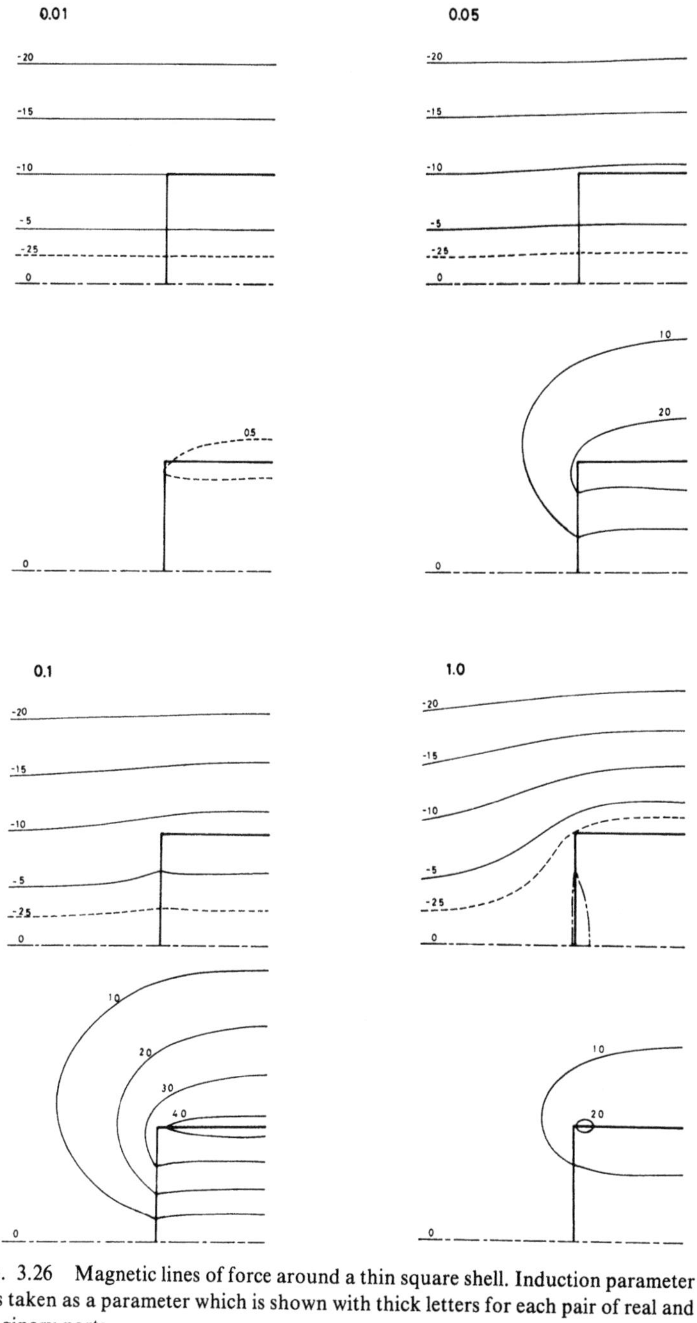

FIG. 3.26 Magnetic lines of force around a thin square shell. Induction parameter Ω is taken as a parameter which is shown with thick letters for each pair of real and imaginary parts.

larger radius. Although it is desirable to make use of copper, alminum and the like for this purpose, a cylindrical iron shell cut from a used oxygen cylinder is often used for economic reasons.

There are two possible ways to arrange such shielding. We may first consider electromagnetic shielding by a conducting shell and then introduce magnetic shielding by a highly-permeable shell in the space already surrounded by the conductor. We may also think of an arrangement in which the order of electromagnetic and magnetic shielding is reversed. Since it is not entirely clear from intuition which is better for achieving an efficient shielding, evaluation of the shielding power will be made for both the cases in this section.

Let us consider a system composed of a thin conducting spherical shell having an integrated conductivity K and a spherical shell of ferromagnetic material having magnetic permeability μ. $\mu=1$ is assumed for the former shell, while zero conductivity is assumed for the latter. The conducting shell may be assumed to be infinitesimally thin for the sake of mathematical convenience because such a thin sheet seems to provide a good approximation to a shell whose finite thickness is relatively small, as we have seen in Subsection 3.6.4.

In Model 1 in Fig. 3.27, a thin conducting shell of radius R_1 is underlain by a highly-permeable shell of which the outer and inner radii are R_1 and R_2, respectively. Meanwhile, a highly-permeable shell of the

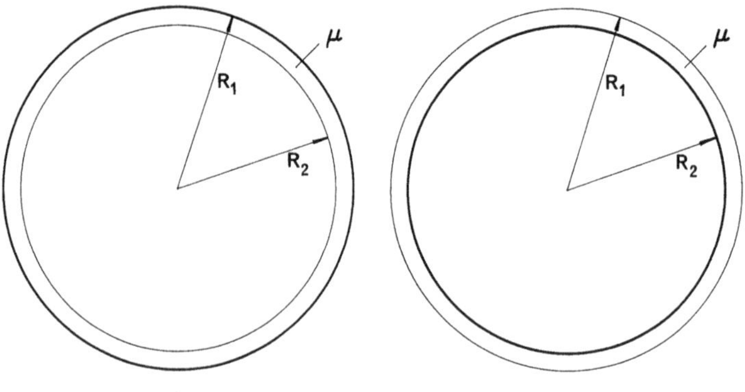

FIG. 3.27 Combinations of a thin electrically conducting sheet and magnetically permeable shell having outer and inner radii of R_1 and R_2, respectively. Model 1 shows the case for which the conducting sheet lies outside the permeable shell of magnetic permeability μ. For Model 2 the conducting sheet is surrounded by the ferromagnetic shell.

same size overlies a thin conducting shell having a radius R_2 in Model 2.

Let us assume that q represents the ratio of the inner radius to the outer one, so that

$$q=R_2/R_1. \tag{3.196}$$

It is further assumed that the integrated conductivity K is given by

$$K=\sigma L$$

where σ is the specific conductivity and L the thickness of the sheet. As an example, we take

$$L=R_1-R_2$$

so that

$$K=\sigma(1-q)\,R_1. \tag{3.197}$$

Returning to the equations in (2.7), it is clear that the magnetic potentials in the regions respectively defined by $R>R_1$, $R_1>R>R_2$ and $R_2>R$ are given by

$$\begin{aligned} W_0&=\left(-H_0R+\frac{I_0}{R^2}\right)\cos\theta \qquad R>R_1\\ W_1&=\left(-H_1R+\frac{I_1}{R^2}\right)\cos\theta \qquad R_1>R>R_2 \end{aligned} \tag{3.198}$$

and

$$W_2=-H_2R\cos\theta \qquad R_2>R$$

where a spherical polar coordinate system (R, θ, ϕ) is used. It is assumed here that a uniform magnetic field of intensity H_0 is applied to the system in a direction parallel to the $\theta=0$ axis.

3.7.1 Model 1

The current function for the electric currents induced in the thin conducting shell at $R=R_1$ is assumed to be given by

$$\Psi_1=K_1\cos\theta \tag{3.199}$$

in a manner similar to (3.185).

From (3.188), we see that K_1 is given as

$$2\,K_1=R_1^2\,Kp\left(-H_0-2\,\frac{I_0}{R_1^3}\right) \tag{3.200}$$

where p replaces time operator $\partial/\partial t$ as before.

In a fashion similar to (3.168), the jump in magnetic potential at $R=R_1$ due to the current sheet implies

$$-H_0R_1+\frac{I_0}{R_1^{\,2}}+H_1-\frac{I_1}{R_1^{\,2}}=4\pi K_1, \tag{3.201}$$

which provides one of the boundary conditions at $R=R_1$ (with K_1 given by (3.200)).

For a purely periodic variation, we put

$$p=i\omega, \qquad i=\sqrt{-1}$$

where ω is the angular frequency and we define an induction parameter by

$$\Omega=\sqrt{4\,\pi\sigma\omega}\,R_1.$$

In that case

$$4\pi R_1Kp=i\Omega^2(1-q)=\gamma \tag{3.202}$$

so that (3.201) becomes

$$-H_0(1-\gamma)+\frac{I_0}{R_1^{\,3}}(1+\gamma)=-H_1+\frac{I_1}{R_1^{\,3}}. \tag{3.203}$$

The boundary condition that the normal component of magnetic flux density is continuous at $R=R_1$ and the two conditions at $R=R_2$ are exactly the same as the second equation of (2.9) and those in (2.10), respectively.

Solving the four equations thus obtained, we obtain S_i as defined by (2.16) as

$$S_i=\frac{9\,\mu}{(\mu+2)(2\mu+1)-2(\mu-1)^2q^3+[2\mu+1-2(\mu-1)q^3]\mu\gamma}$$

which can be rewritten as

$$S_i=\frac{1}{1+\frac{2}{9}\frac{(\mu-1)^2}{\mu}(1-q^3)+i\,\frac{\Omega^2(1-q)}{9}\,[2\,\mu+1-2(\mu-1)q^3]}. \tag{3.204}$$

It is obvious that

$$\lim_{\sigma\to 0} S_i=\frac{1}{1+\frac{2}{9}\frac{(\mu-1)^2}{9}(1-q^3)}$$

which agrees with the shielding ratio for the system without the conductor as given by (2.17). Conversely, making $\mu \to 1$, we arrive at

$$\lim_{\mu \to 1} S_i = \frac{1}{1+i\Omega^2(1-q)/3} = \frac{1}{1+i\Omega^2(L/3\,R_1)}$$

which agrees with (3.191), the shielding ratio for the system without the magnetic shell.

When $\sigma \to \infty$, we also see that

$$\lim_{\sigma \to \infty} S_i = 0.$$

We therefore conclude that the shielding power of the system tends to that of the spherical shell having permeability μ as studied in Subsection 2.2.1.1 when $\sigma \to 0$, while the effect of electromagnetic shielding becomes appreciable when σ, and so Ω, take on large values.

Assuming $q=0.9$, variations of mod S_i and arg S_i with Ω are calculated for the two cases $\mu=100$ and 1000. These are illustrated in Figs. 3.28 and 3.29, respectively, with solid lines. In the figures, we see

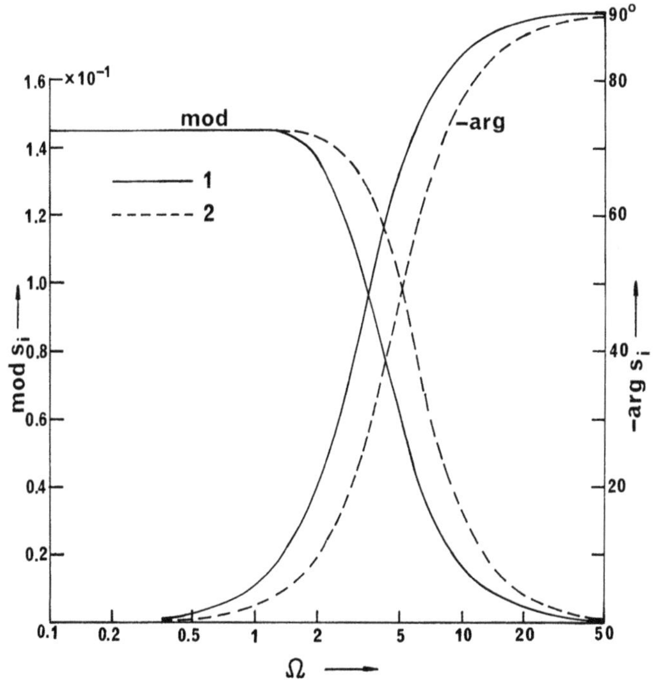

FIG. 3.28 Changes in mod S_i and arg S_i with Ω for Models 1 and 2 as shown by solid and dashed lines, respectively. $q=0.9$ and $\mu=100$ are assumed.

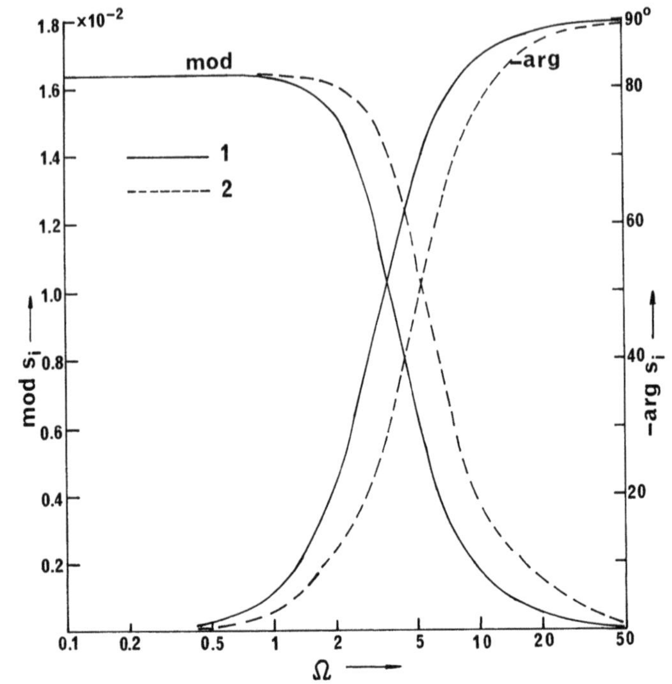

FIG. 3.29 Changes in mod S_i and arg S_i with Ω for Models 1 and 2 as shown by solid and dashed lines, respectively. $q=0.9$ and $\mu=1000$ are assumed.

that the shielding ratio is almost equal to that of the permeable shell for the static case when Ω is smaller than 1. However, the effect of electromagnetic induction becomes appreciable for values of Ω larger than 2, so that the shielding ratio becomes very much smaller especially for Ω's larger than 10.

3.7.2 Model 2

A similar consideration about the potential jump at $R=R_2$ leads to one of the boundary conditions at $R=R_2$ being

$$-H_1+\frac{I_1}{{R_2}^3}=\left(-1-\frac{\gamma q}{2}\right)H_2\,. \tag{3.205}$$

Other conditions at $R=R_1$ and $R=R_2$ are exactly the same as (2.9) and the second equation of (2.10). Solving these equations, we get

$$S_i=\frac{9\,\mu}{(\mu+2)(2\,\mu+1)-2(\mu-1)^2q^3+[\mu+2-(\mu-1)q^3]\mu\gamma q}$$

which can be rewritten as

$$S_i = \frac{1}{1+\frac{2}{9}\frac{(\mu-1)^2}{\mu}(1-q^3)+i\frac{\Omega^2(1-q)}{9}[\mu+2-(\mu-1)\,q^3]\,q}\,. \tag{3.206}$$

Corresponding to the two extreme cases $\sigma\to 0$ and $\sigma\to\infty$, we reach the same conclusions as mentioned in the case of Model 1. When $\mu\to 1$, we obtain

$$\lim_{\mu\to 1} S_i = \frac{1}{1+i\Omega_q{}^2(1-q)/3q} = \frac{1}{1+i\Omega_q{}^2(L/3R_2)}$$

where

$$\Omega_q = q\,\Omega\,.$$

This is what we expect for the thin shell placed in free space. Mod S_i and arg S_i of Model 2 for $\mu=100$ and 1000 are shown by dashed lines in Figs. 3.28 and 3.29.

3.7.3 Comparison of shielding power between Models 1 and 2

Looking at Figs. 3.28 and 3.29, we see that the shielding ratio for Model 1 is smaller than that for Model 2. This means that the shielding power of Model 1 is higher than that of Model 2. Part of this difference in the shielding power is due to the fact that the radius of the conducting shell of Model 1 is larger than that of Model 2.

In order to compare the shielding power in more detail, ratios of mod S_i for Model 1 to that for Model 2 are calculated for $\mu=100$ and 1000 as shown in Fig. 3.30. We see that there is almost no difference in the shielding power for Ω equal to or smaller than 2. However, the ratio becomes as small as 0.47–0.45 when Ω exceeds 10 or so. A fairly steep change in the ratio of the shielding power takes place between $\Omega=2$ and 5.

It may therefore be said that it is better to use a conducting shell surrounding a highly-permeable shell in order to achieve the best shielding to an external field. The degree of shielding to be achieved can be inferred from the graphs in Figs. 3.28 and 3.29.

Let us assume $\sigma=10^8$ ohm^{-1} m^{-1}, $R_1=10$ cm and $q=0.9$, Ω is then 2.8, 8.9 and 19.9 for frequencies of 1, 10 and 50 Hz, respectively. It is then seen that a uniform external magnetic field of frequency 50 Hz is reduced to about one-hundredth and one-thousandth of its value by Model 1 shells having magnetic permeabilities 100 and 1000, respectively.

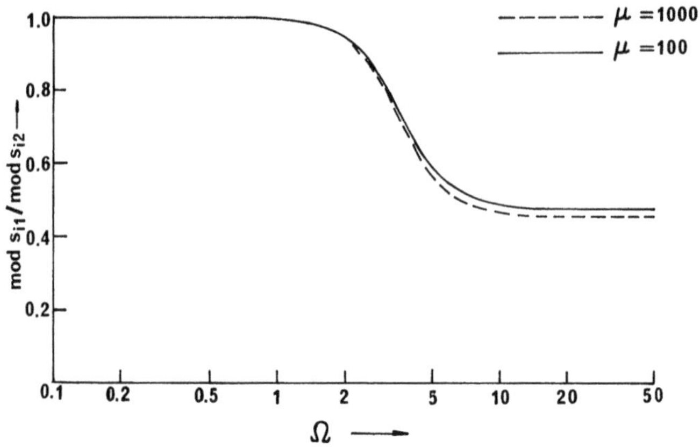

FIG. 3.30 Ratios of mod S_i for Model 1 to that for Model 2 as a function of Ω.

The above conclusion concerning the shielding power of a conducting shell combined with a highly-permeable shell may also hold good approximately for cylindrical shells judging from the similarity between spherical and cylindrical shells as discussed in the foregoing sections.

In order to achieve efficient shielding to an oscillating magnetic field with a low frequency, one should note the above conclusions.

3.8 Two-Layer Conductor

There are a number of problems of electromagnetic induction in a system of conductors in which a superficial layer is underlain by another layer having a different conductivity. As one of the typical problems of this kind, it is inferred that the earth's superficial layer including the ocean is underlain by a layer of low conductivity, although a much more conducting layer is supposed to exist at greater depth (RIKITAKE, 1966).

As a logical extension of the studies on the shielding power of conducting shells as developed in the foregoing sections of this chapter, it is planned in this section to deal with the shielding power of a two-layer conductor.

3.8.1 Electromagnetic coupling between superficial and deeper conductors

Probably, RIKITAKE (1961) was the first who pointed out the

importance of electromagnetic coupling between a superficial conducting layer and a deep-seated one when studying electromagnetic induction problems such as the effect of a vast ocean on geomagnetic variations of short period.

Let us study a spherical conductor, for which the conductivity is infinitely large, covered by a concentric thin spherical sheet having a uniform integrated conductivity K. The radii of these conductors are respectively R_2 and R_1 as shown in Fig. 3.31.

When an oscillating magnetic field of which the typical magnetic potential is defined by $R_1 e_n{}^m (R/R_1)^n S_n{}^m$ is applied to these conductors, the current function for the electric currents induced in the sheet may be written as

$$\Psi_1 = a_n{}^m \, S_n{}^m \qquad (R=R_1). \tag{3.207}$$

We also can define a current function at $R=R_2$ as

$$\Psi_2 = b_n{}^m \, S_n{}^m \tag{3.208}$$

because of the assumption of perfect conductivity for the inner sphere. $S_n{}^m$ stands for a surface harmonic function $P_n{}^m(\cos\theta) \, {}^{\cos}_{\sin} \, m\phi$ as before.

The magnetic potentials due to the electric currents defined by these current functions may be readily obtained as

$$\begin{aligned} w_1 &= 4\pi \frac{n}{2n+1} a_n{}^m \left(\frac{R}{R_1}\right)^{-n-1} S_n{}^m \qquad (R>R_1) \\ w_1' &= -4\pi \frac{n+1}{2n+1} a_n{}^m \left(\frac{R}{R_1}\right)^{n} S_n{}^m \qquad (R_1>R) \end{aligned} \tag{3.209}$$

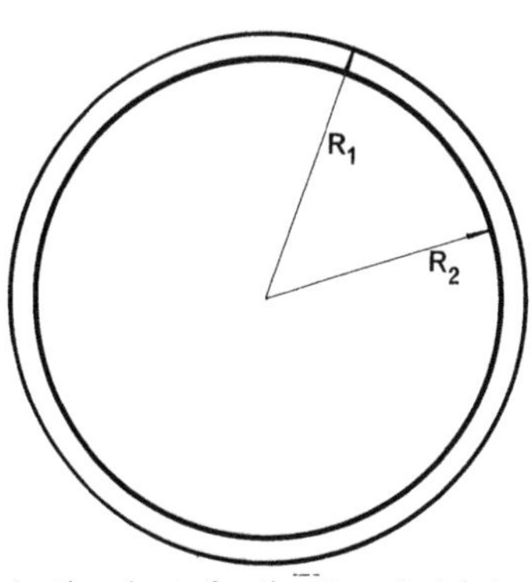

FIG. 3.31 A thin conducting sheet of radius R_1 underlain by a perfectly-conducting sphere of radius R_2.

and

$$
\begin{aligned}
w_2 &= 4\pi \frac{n}{2n+1} b_n^m \left(\frac{R}{R_2}\right)^{-n-1} S_n^m \qquad (R>R_2) \\
w_2' &= -4\pi \frac{n+1}{2n+1} b_n^m \left(\frac{R}{R_2}\right)^{n} S_n^m \qquad (R_2>R),
\end{aligned}
\tag{3.210}
$$

respectively.

We then obtain the magnetic potentials for regions $R>R_1$ and $R_1>R>R_2$ as

$$
\begin{aligned}
W_n^m &= \left[R_1 e_n^m \left(\frac{R}{R_1}\right)^n + 4\pi \frac{n}{2n+1} a_n^m \left(\frac{R}{R_1}\right)^{-n-1} \right. \\
&\left. + 4\pi \frac{n}{2n+1} b_n^m \left(\frac{R}{R_2}\right)^{-n-1}\right] S_n^m \qquad (R>R_1) \\
W_n^{m\prime} &= \left[R_1 e_n^m \left(\frac{R}{R_1}\right)^n - 4\pi \frac{n+1}{2n+1} a_n^m \left(\frac{R}{R_1}\right)^{n} \right. \\
&\left. + 4\pi \frac{n}{2n+1} b_n^m \left(\frac{R}{R_2}\right)^{-n-1}\right] S_n^m \qquad (R_1>R>R_2).
\end{aligned}
\tag{3.211}
$$

The condition that the normal component of magnetic field should vanish at the surface of the perfect conductor, where $R=R_2$, implies

$$
n e_n^m q^{n-1} - \frac{4\pi}{R_2} \frac{n(n+1)}{2n+1} (a_n^m q^n + b_n^m) = 0 \tag{3.212}
$$

where

$$
q = R_2/R_1 .
$$

The boundary condition for the thin sheet at $R=R_1$ which is similar to (3.182) leads to

$$
n(n+1) a_n^m = R_1^2 Kp \left[n e_n^m - \frac{4\pi}{R_1} \frac{n(n+1)}{2n+1} (a_n^m + b_n^m q^{n+1}) \right] \tag{3.213}
$$

where p replaces the time operator $\partial/\partial t$.

Solving (3.212) and (3.213), we obtain

$$
\frac{a_n^m}{e_n^m} = \frac{R_1}{4\pi} \frac{2n+1}{n+1} \frac{(1-q^{2n+1})\,T}{2n+1+(1-q^{2n+1})\,T} \tag{3.214}
$$

$$
\frac{b_n^m}{e_n^m} = \frac{R_1}{4\pi} \frac{(2n+1)^2}{n+1} \frac{q^n}{2n+1+(1-q^{2n+1})\,T} \tag{3.215}
$$

where

$$T=4\pi R_1 Kp.$$

When the temporal variation is proportional to $e^{i\omega t}$, we put $p=i\omega$ and obtain

$$T=4\pi\sigma LR_1 i\omega$$

which is rewritten as

$$T=i\Omega^2(L/R_1)$$

where, denoting the specific conductivity and the actual thickness of the sheet by σ and L, we define

$$K=\sigma L$$

and

$$\Omega=\sqrt{4\pi\sigma\omega}\, R_1.$$

We thus rewrite (3.214) and (3.215) as

$$\frac{a_n{}^m}{e_n{}^m}=\frac{R_1}{4\pi}\frac{2n+1}{n+1}\frac{(1-q^{2n+1})i\Omega^2(L/R_1)}{2n+1+(1-q^{2n+1})i\Omega^2(L/R_1)} \tag{3.216}$$

and

$$\frac{b_n{}^m}{e_n{}^m}=\frac{R_1}{4\pi}\frac{(2n+1)^2}{n+1}\frac{q^n}{2n+1+(1-q^{2n+1})i\Omega^2(L/R_1)}. \tag{3.217}$$

When there is no inner conductor, we put $q\rightarrow 0$. In that case,

$$\begin{aligned}\left(\frac{a_n{}^m}{e_n{}^m}\right)_{q=0}&=\frac{R_1}{4\pi}\frac{2n+1}{n+1}\frac{i\Omega^2(L/R_1)}{2n+1+i\Omega^2(L/R_1)}\\ \left(\frac{b_n{}^m}{e_n{}^m}\right)_{q=0}&=0.\end{aligned} \tag{3.218}$$

From the second equation of (3.211), we then obtain

$$W_n{}^{m\prime}=R_1\left(\frac{R}{R_1}\right)^n\frac{2n+1}{2n+1+i\Omega^2(L/R_1)}. \tag{3.219}$$

We therefore see that the magnetic potential inside the sheet becomes smaller than that of the inducing field by a factor $(2n+1)/\{2n+1+i\Omega^2(L/R_1)\}$. This conclusion agrees with the former one for a thin sheet as given by (3.190).

We also have

$$\frac{a_n^{\ m}}{a_{n,0}^{\ m}}=\frac{2n+1+i\Omega^2(L/R_1)}{(2n+1)/(1-q^{2n+1})+i\Omega^2(L/R_1)}. \tag{3.220}$$

As $q<1$, the modulus of the denominator is always larger than that of the numerator, so that the current function in the sheet is reduced by electromagnetic coupling between the sheet and the underlying conductor. Assuming $L/R_1=0.2$ and $n=1$, changes in the modulus of the ratio with the induction parameter Ω are calculated for a few values of q as shown in Fig. 3.32. $m=0$ is assumed, too.

We can clearly see in the figure that mod $a_1^{\ 0}/a_{1,0}^{\ 0}$ approaches 1 asymptotically as Ω becomes larger. This means that the sheet behaves like a perfect conductor for a high frequency field, so that the influence of the deeper layer is small. On the contrary, mod $a_1/a_{1,0}^{\ 0}$ tends to $1-q^3$ for smaller values of Ω. In that case the reduction of the current intensity due to electromagnetic coupling between the two conductors becomes appreciable when q takes on a value around 1 or smaller. The degree of reduction rapidly changes as Ω changes from 2 to 10.

We are now in a position to compare the intensity of the magnetic field immediately inside the sheet to that of the applied field. Differentiating the second equation of (3.211) with respect to R and θ, the normal and tangential components at $R=R_1$ are obtained as

$$H_R=\left[n\,e_n^{\ m}-\frac{4\pi}{R_1}\frac{n(n+1)}{2n+1}a_n^{\ m}-\frac{4\pi}{R_1}\frac{n(n+1)}{2n+1}q^{n+1}\,b_n^{\ m}\right]S_n^{\ m}$$

and (3.221)

$$H_\theta=\left[e_n^{\ m}-\frac{4\pi}{R_1}\frac{n+1}{2n+1}a_n^{\ m}+\frac{4\pi}{R_1}\frac{n}{2n+1}q^{n+1}b_n^{\ m}\right]\frac{\partial S_n^{\ m}}{\partial\theta},$$

respectively.

With the aid of (3.216) and (3.217), $a_n^{\ m}$ and $b_n^{\ m}$ can be eliminated from (3.221) to obtain

$$H_R=\frac{n(2n+1)(1-q^{2n+1})}{2n+1+(1-q^{2n+1})\,i\Omega^2(L/R_1)}e_n^{\ m}S_n^{\ m} \tag{3.222}$$

and

$$H_\theta=\frac{(2n+1)(1+nq^{2n+1})}{2n+1+(1-q^{2n+1})\,i\Omega^2(L/R_1)}e_n^{\ m}\frac{\partial S_n^{\ m}}{\partial\theta}. \tag{3.223}$$

We therefore see that the amplitudes of the normal and tangential components of magnetic field are modified by factors

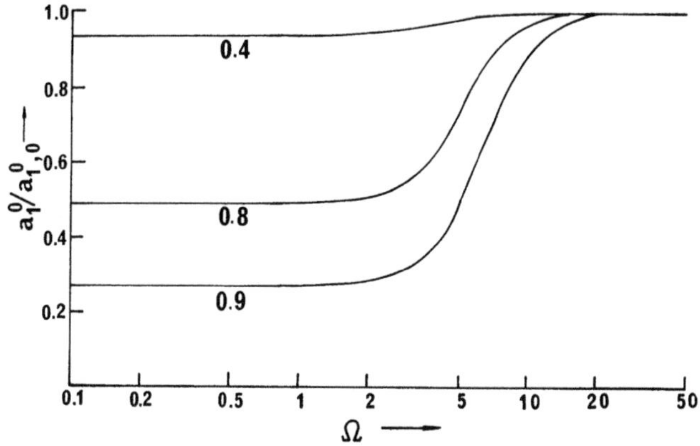

FIG. 3.32 Changes in mod $a_1^0/a_{1,0}^0$, with Ω. $L/R=0.2$ is assumed. Three values 0.4, 0.8 and 0.9 are assigned to q.

$$\frac{n(2n+1)(1-q^{2n+1})}{2n+1+(1-q^{2n+1})i\Omega^2(L/R_1)}$$

and

$$\frac{(2n+1)(1+nq^{2n+1})}{2n+1+(1-q^{2n+1})i\Omega^2(L/R_1)},$$

respectively.

When the inner conductor is absent, the magnetic field components immediately inside the sheet are obtained as

$$H_{R,0}=\frac{n(2n+1)}{2n+1+i\,\Omega^2(L/R_1)}e_n{}^m S_n{}^m \tag{3.224}$$

and

$$H_{\theta,0}=\frac{2n+1}{2n+1+i\Omega^2(L/R_1)}\,e_n{}^m\,\frac{\partial S_n{}^m}{\partial\theta} \tag{3.225}$$

so that we obtain

$$\begin{aligned}\frac{H_R}{H_{R,0}}&=(1-q^{2n+1})\frac{2n+1+i\Omega^2(L/R_1)}{2n+1+(1-q^{2n+1})i\Omega^2(L/R_1)}\\ \frac{H_\theta}{H_{\theta,0}}&=(1+nq^{2n+1})\frac{2n+1+i\Omega^2(L/R_1)}{2n+1+(1-q^{2n+1})i\Omega^2(L/R_1)}.\end{aligned} \tag{3.226}$$

Thus the effect of electromagnetic coupling between the sheet and the inner conductor on the amplitude of the normal magnetic field is different from the effect on the tangential component.

When $n=1$, the total field intensity F is given by

$$F=\frac{3}{3+i(1-q^3)\Omega^2(L/R_1)}[(1-q^3)^2\cos^2\theta+(1+q^3)^2\sin^2\theta]^{1/2}e_1{}^0.$$

Denoting the total intensity when there is no inner conductor by F_0, we have

$$\frac{F}{F_0}=\frac{3+i\Omega^2(L/R_1)}{3+i(1-q^3)\Omega^2(L/R_1)}\left(\frac{F}{F_0}\right)_\theta$$
$$\left(\frac{F}{F_0}\right)_\theta=[(1-q^3)^2\cos^2\theta+(1+q^3)^2\sin^2\theta]^{1/2} \tag{3.227}$$

from which it is clear that the phase of F is different from that of F_0.

Changes in $(F/F_0)_\theta$ with θ are shown in Fig. 3.33 taking q as a parameter. The magnetic field applied to the conductors from outside is assumed to be parallel to the $\theta=0$ axis. When the radius of the inner conductor is large, the ratio $(F/F_0)_\theta$ becomes small for small values of θ. In this case the cancelling effect of the inner conductor is remarkable. When θ approaches $\pi/2$, however, the total field intensity becomes larger because the magnetic lines of force tend to be deflected from the inner conductor and so increasing the magnetic flux density in the

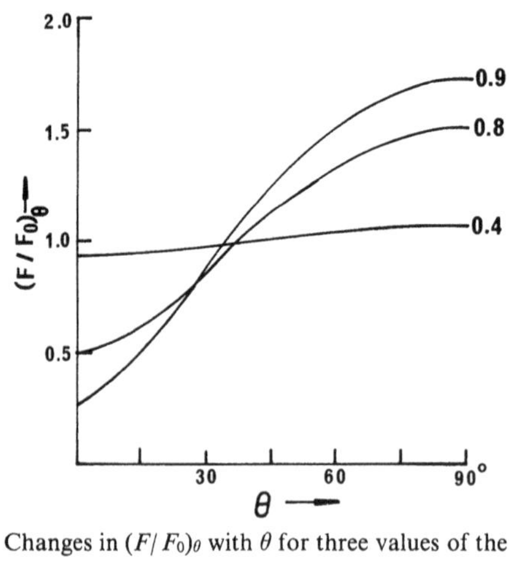

FIG. 3.33 Changes in $(F/F_0)_\theta$ with θ for three values of the parameter q.

space between $R=R_1$ and $R=R_2$. In this case, therefore, electromagnetic coupling acts in such a way as to increase the magnetic field intensity. When the radius of the inner sphere is smaller than $0.4R_1$, the influence of the inner sphere on $(F/F_0)_\theta$ becomes negligible, so that the shielding effect of the sheet tends to be independent of θ.

When Ω tends to 0, we see that $\{3+i\Omega^2(L/R_1)\}/\{3+i(1-q^3)\Omega^2(L/R_1)\}\to 1$. In that case we have $F/F_0\to(F/F_0)_\theta$. Figure 3.33 therefore illustrates how F/F_0 depends on θ for a few values of q when the frequency is very low. On the other hand, the factor to be multiplied by $(F/F_0)_\theta$ tends to $(1-q^3)^{-1}$ when $\Omega\to\infty$. In this case, the values expressed by the curves in Fig. 3.33 should be multiplied by 3.69, 2.05 and 1.07 respectively for $q=0.9$, 0.8 and 0.4 in order to obtain F/F_0. It is therefore seen that the intensification of the magnetic field for θ close to $\pi/2$ becomes appreciable for higher frequencies.

It was demonstrated in this subsection that electromagnetic coupling between a superficial layer and a deep-seated layer is sometimes very important when discussing electromagnetic shielding, especially when the inner layer is placed close to the outer layer.

3.8.2 Plane layer model

Let us take a two-layer model as shown in Fig. 3.34, which is similar to Fig. 3.1, but where the lower non-conducting layer is here replaced by a layer of conductivity σ_2. The overlying conducting layer is assumed to have conductivity σ_1. For the sake of simplicity, the

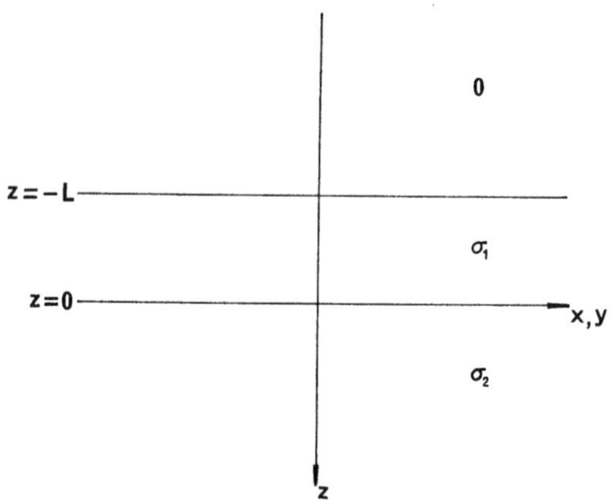

FIG. 3.34 Two-layer plane model. The conductivities for $0>z>-L$ and $z>0$ are σ_1 and σ_2, respectively. The space for $-L>z$ is non-conducting.

magnetic permeability is assumed unity everwhere, in electromagnetic units.

The theory of electromagnetic induction in such a two-layer model can be presented in a fashion similar to that of Section 3.2. In the non-conducting region defined by $z<-L$, we have a magnetic potential which satisfies (3.18). The typical term of magnetic potential is given by

$$W_{mn}=(e_{mn}\,e^{-kz}+i_{mn}\,e^{kz})\,V_{mn} \tag{3.228}$$

which is the same as (3.19). k is defined by (3.20), and a harmonic function V_{mn} satisfies Eq. (3.21).

The x, y and z components of the magnetic field for $z<-L$ are then obtained as

$$\boldsymbol{H}=\begin{cases} -(e_{mn}\,e^{-kz}+i_{mn}\,e^{kz})\dfrac{\partial V_{mn}}{\partial x} \\ -(e_{mn}\,e^{-kz}+i_{mn}\,e^{kz})\dfrac{\partial V_{mn}}{\partial y} \qquad (z<-L). \\ -k(-e_{mn}\,e^{-kz}+i_{mn}\,e^{kz})\,V_{mn} \end{cases} \tag{3.229}$$

In the conducting regions, we difine vector potentials $\boldsymbol{A}_1$ and $\boldsymbol{A}_2$ which respectively satisfy

$$\begin{aligned} &(\nabla^2-\chi_1^{\,2})\boldsymbol{A}_1=0 \qquad (-L<z<0) \\ &(\nabla^2-\chi_2^{\,2})\,\boldsymbol{A}_2=0 \qquad (0<z) \end{aligned} \tag{3.230}$$

where

$$\chi_1^{\,2}=4\pi\sigma_1 p, \qquad \chi_2^{\,2}=4\pi\sigma_2 p \tag{3.231}$$

in which p replaces the time operator $\partial/\partial t$.

Typical solutions of the two equations in (3.230) can be given as

$$\boldsymbol{A}_{1,mn}=(C_{mn}e^{-\sqrt{k^2+\chi_1^{\,2}}\,z}+D_{mn}\,e^{\sqrt{k^2+\chi_1^{\,2}}\,z})(\boldsymbol{r}\wedge\text{grad}\ V_{mn}) \qquad (-L<z<0) \tag{3.232}$$

and

$$\boldsymbol{A}_{2,mn}=E_{mn}\,e^{-\sqrt{k^2+\chi_2^{\,2}}\,z}\,(\boldsymbol{r}\wedge\text{grad}\ V_{mn}) \quad (0<z) \tag{3.233}$$

in which $(\boldsymbol{r}\wedge\text{grad}\ V_{mn})$ has been defined in (3.28). No term proportional to $e^{\sqrt{k^2+\chi_2^2}z}$ is included in (3.233) because the vector potential should not become infinitely large as $z\to\infty$ when we consider electromagnetic induction by a magnetic field arising from a source in $z<-L$.

We then obtain, from (3.12), the magnetic field components as

$$H=\begin{cases}(-C_{mn}\,e^{-\sqrt{k^2+\chi_1^2}\,z}+D_{mn}\,e^{\sqrt{k^2+\chi_1^2}\,z})\sqrt{k^2+\chi_1^2}\,\dfrac{\partial V_{mn}}{\partial x} \\ (-C_{mn}\,e^{-\sqrt{k^2+\chi_1^2}\,z}+D_{mn}\,e^{\sqrt{k^2+\chi_1^2}\,z})\sqrt{k^2+\chi_1^2}\,\dfrac{\partial V_{mn}}{\partial y} & (-L<z<0) \\ (C_{mn}\,e^{-\sqrt{k^2+\chi_1^2}\,z}+D_{mn}\,e^{\sqrt{k^2+\chi_1^2}\,z})\,k^2\,V_{mn}\end{cases} \tag{3.234}$$

and

$$H=\begin{cases}-E_{mn}\,e^{-\sqrt{k^2+\chi_2^2}\,z}\sqrt{k^2+\chi_2^2}\,\dfrac{\partial V_{mn}}{\partial x} \\ -E_{mn}\,e^{-\sqrt{k^2+\chi_2^2}\,z}\sqrt{k^2+\chi_2^2}\,\dfrac{\partial V_{mn}}{\partial y} & (0<z). \\ E_{mn}\,e^{-\sqrt{k^2+\chi_2^2}\,z}\,k^2 V_{mn}\end{cases} \tag{3.235}$$

The continuity conditions of the magnetic field components at $z=-L$ and $z=0$ imply

$$\begin{aligned}&e_{mn}\,e^{kL}+i_{mn}\,e^{-kL}=\sqrt{k^2+\chi_1^2}\,(C_{mn}\,e^{\sqrt{k^2+\chi_1^2}\,L}-D_{mn}\,e^{-\sqrt{k^2+\chi_1^2}\,L})\\ &e_{mn}\,e^{kL}-i_{mn}\,e^{-kL}=k(C_{mn}\,e^{\sqrt{k^2+\chi_1^2}\,L}+D_{mn}\,e^{-\sqrt{k^2+\chi_1^2}\,L})\\ &\sqrt{k^2+\chi_1^2}\,(C_{mn}-D_{mn})=\sqrt{k^2+\chi_2^2}\,E_{mn}\\ &C_{mn}+D_{mn}=E_{mn}.\end{aligned} \tag{3.236}$$

Solving (3.236), we obtain

$$E/e=4\sqrt{k^2+\chi_1^2}\,e^{kL}/\Phi_1 \tag{3.237}$$

$$\begin{aligned}i/e=e^{2kL}\,[&(\sqrt{k^2+\chi_1^2}+\sqrt{k^2+\chi_2^2})(\sqrt{k^2+\chi_1^2}-k)\,e^{\sqrt{k^2+\chi_1^2}L}\\ &-(\sqrt{k^2+\chi_1^2}-\sqrt{k^2+\chi_2^2})(\sqrt{k^2+\chi_1^2}+k)\,e^{-\sqrt{k^2+\chi_1^2}L}]/\Phi_1\end{aligned} \tag{3.238}$$

$$\begin{aligned}\Phi_1=&(\sqrt{k^2+\chi_1^2}+\sqrt{k^2+\chi_2^2})(\sqrt{k^2+\chi_1^2}+k)\,e^{\sqrt{k^2+\chi_1^2}L}\\ &-(\sqrt{k^2+\chi_1^2}-\sqrt{k^2+\chi_2^2})(\sqrt{k^2+\chi_1^2}-k)\,e^{-\sqrt{k^2+\chi_1^2}L}.\end{aligned} \tag{3.239}$$

In (3.237), (3.238) and (3.239), suffixes m and n are dropped for the sake of simplicity. Coefficients C and D are also readily obtained although they are not given here.

For a periodic variation for which the time factor is given by $e^{i\omega t}$, we define induction parameters as

$$\Omega_1=\sqrt{4\pi\sigma_1\omega}\,L,\qquad \Omega_2=\sqrt{4\pi\sigma_2\omega}\,L.$$

From (3.231), we obtain

$$\chi_1^2=i(\Omega_1/L)^2,\quad \chi_2^2=i(\Omega_2/L)^2.$$

We further define

$$\left.\begin{aligned}\lambda_1=\chi_1/k=\frac{1+i}{\sqrt{2}}\,\Omega_1/kL\\ \lambda_2=\chi_2/k=\frac{1+i}{\sqrt{2}}\,\Omega_2/kL\end{aligned}\right\}. \tag{3.240}$$

With the aid of these relations, (3.237), (3.238) and (3.239) can be written as

$$E/e=4\sqrt{1+\lambda_1^2}\,e^{kL}\,\Phi_1' \tag{3.241}$$

$$\begin{aligned}i/e=[&(\sqrt{1+\lambda_1^2}+\sqrt{1+\lambda_2^2})(\sqrt{1+\lambda_1^2}-1)\,e^{kL\sqrt{1+\lambda_1^2}}\\ &-(\sqrt{1+\lambda_1^2}-\sqrt{1+\lambda_2^2})(\sqrt{1+\lambda_1^2}+1)\,e^{-kL\sqrt{1+\lambda_1^2}}]\,e^{2\,kL}/\Phi_1'\end{aligned} \tag{3.242}$$

and

$$\begin{aligned}\Phi_1'=&(\sqrt{1+\lambda_1^2}+\sqrt{1+\lambda_2^2})(\sqrt{1+\lambda_1^2}+1)\,e^{kL\sqrt{1+\lambda_1^2}}\\ &-(\sqrt{1+\lambda_1^2}-\sqrt{1+\lambda_2^2})(\sqrt{1+\lambda_1^2}-1)\,e^{-kL\sqrt{1+\lambda_1^2}}.\end{aligned} \tag{3.243}$$

As the x and z components of the inducing magnetic field at $z=0$ are respectively given by

$$H_{x0}=-e\frac{\partial V}{\partial x}, \qquad H_{z0}=k\,e\,V,$$

the ratios of the magnetic field components immediately beneath the overlying layer to those of the inducing field are given by

$$\begin{aligned}R_x=(H_x/H_{x0})_{z=0}=4\sqrt{1+\lambda_1^2}\sqrt{1+\lambda_2^2}\,e^{kL}/\Phi_1'\\ R_z=(H_z/H_{z0})_{z=0}=4\sqrt{1+\lambda_1^2}\,e^{kL}/\Phi_1'.\end{aligned} \tag{3.244}$$

Figure 3.35 shows the changes in mod R_x and mod R_z with Ω_1 when $\Omega_2=100\ \Omega_1$ and $kL=0.1$. As σ_2 is equal to 10000 σ_1, this case provides a model in which a moderately conducting layer is underlain by a highly conducting semi-infinite medium. Comparing Fig. 3.35 to Fig. 3.2, we notice that the underlying layer cancels the magnetic field normal to the boundaries to a large extent. Conversely, the tangential magnetic field is enhanced by the presence of the underlying layer by a factor close to 2 for Ω_1 in the range 0.01 to 1 although the effect vanishes for larger values of Ω_1.

Changes in the argument of R_x and R_z are complicated. Since the phase relation is seldom of concern in actual problems, no illustrations of the argument are presented here.

The electromagnetic response of the system of conductors as defined by (3.242) is also not given here because we are mostly

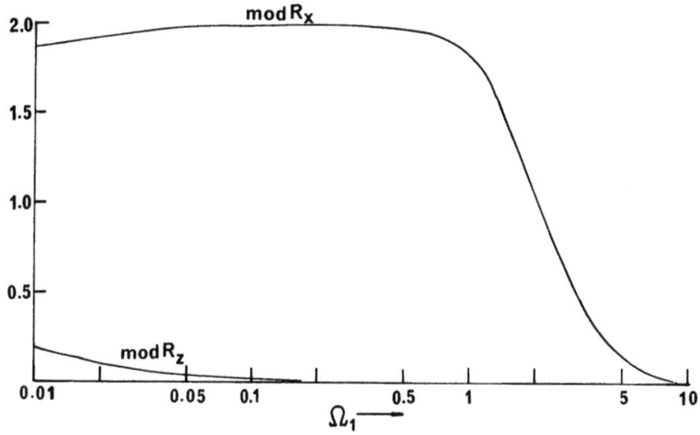

FIG. 3.35 Changes in mod R_x and mod R_z with Ω_1 when $\Omega_2 = 100\ \Omega_1$ and $kL = 0.1$. The definitions of R_x and R_z are given by the equations in (3.244).

concerned with the shielding effect.

What we have studied in this and preceding subsections suggests that the influence of a conductor placed in the space surrounded by a conducting layer is important when discussing the shielding effect of the latter. Such an inner conductor should not be placed close to the layer when trying to achieve an efficient electromagnetic shielding, otherwise some enhancement of the magnetic field will take place contrary to the desire that we should reduce the magnetic field inside the layer.

An additional account of electromagnetic shielding by a plane slab will be presented in Appendix 2 when the source field arises from within the underlying conductor.

3.8.3 Spherical layer model

This subsection is reserved for two-layer spherical conductors. The configuration of the model is similar to that shown in Fig. 3.31 in which the conductivities are taken as σ_1 and σ_2 respectively for regions $R_1 > R > R_2$ and $R_2 > R$. The theory of electromagnetic induction in such a model can be developed in a fashion similar to that for a spherical shell as given in Section 3.3 although the mathematics inevitably becomes a little more complicated.

Let us attach subscripts 1 and 2 respectively to quantities in regions $R_1 > R > R_2$ and $R_2 > R$. It is obvious that typical solutions of the type given in (3.62) also hold good for the present case. We therefore obtain

$$\boldsymbol{A}_{1,n}=R_1(\boldsymbol{r}\wedge\text{grad } S_n)\, f_{1,n}(p,\rho) \quad (R_1>R>R_2) \tag{3.245}$$

$$\boldsymbol{A}_{2,n}=R_1(\boldsymbol{r}\wedge\text{grad } S_n)\, f_{2,n}(p,\rho) \quad (R_2>R) \tag{3.246}$$

where r, S_n, p and ρ are defined in Subsection 3.3.1. $f_{1,n}$ and $f_{2,n}$ must satisfy differential equations similar to (3.61).

In that case, the components of the magnetic flux density are obtained as

$$\left.\begin{aligned} B_{1,R}&=-\rho^{-1}n(n+1)\, f_{1,n}\, S_n\\ B_{1,\theta}&=-\rho^{-1}\frac{\mathrm{d}(\rho f_{1,n})}{\mathrm{d}\rho}\frac{\partial S_n}{\partial\theta}\\ B_{1,\phi}&=-\rho^{-1}\frac{\mathrm{d}(\rho f_{1,n})}{\mathrm{d}\rho}\frac{\partial S_n}{\sin\theta\,\partial\phi}\end{aligned}\right\} \quad (R_1>R>R_2) \tag{3.247}$$

and

$$\left.\begin{aligned} B_{2,R}&=-\rho^{-1}n(n+1)\, f_{2,n}\, S_n\\ B_{2,\theta}&=-\rho^{-1}\frac{\mathrm{d}(\rho f_{2,n})}{\mathrm{d}\rho}\frac{\partial S_n}{\partial\theta}\\ B_{2,\phi}&=-\rho^{-1}\frac{\mathrm{d}(\rho f_{2,n})}{\mathrm{d}\rho}\frac{\partial S_n}{\sin\theta\,\partial\phi}\end{aligned}\right\} \quad (R_2>R). \tag{3.248}$$

It is obvious that we can define a magnetic potential W, which satisfies $\nabla^2 W=0$ in free space, for $R>R_1$. A typical solution for magnetic potential is then given by

$$W_n=R_1(e_n\rho^n+i_n\rho^{-n-1})S_n \qquad (R>R_1) \tag{3.249}$$

from which we obtain the magnetic field components as

$$\left.\begin{aligned} H_R&=-[ne_n\rho^{n-1}-(n+1)i_n\rho^{-n-2}]S_n\\ H_\theta&=-(e_n\rho^{n-1}+i_n\rho^{-n-2})\frac{\partial S_n}{\partial\theta}\\ H_\phi&=-(e_n\rho^{n-1}+i_n\rho^{-n-2})\frac{\partial S_n}{\sin\theta\,\partial\phi}\end{aligned}\right\} \quad (R>R_1). \tag{3.250}$$

The continuity of the tangential magnetic field and the normal magnetic flux density at $R=R_1$ and $R=R_2$ imply

$$\begin{aligned} ne_n-(n+1)i_n&=n(n+1)(f_{1,n})_{\rho=1}\\ e_n+i_n&=\mu_1^{-1}\left[\frac{\mathrm{d}(\rho f_{1,n})}{\mathrm{d}\rho}\right]_{\rho=1}\end{aligned} \tag{3.251}$$

$$\mu_1^{-1}\left[\frac{\mathrm{d}(\rho f_{1,n})}{\mathrm{d}\rho}\right]_{\rho=q}=\mu_2^{-1}\left[\frac{\mathrm{d}(\rho f_{2,n})}{\mathrm{d}\rho}\right]_{\rho=q}$$
$$(f_{1,n})_{\rho=q}=(f_{2,n})_{\rho=q}$$

where q is given by (3.71). The magnetic permeabilities are assumed to be μ_1 and μ_2 for $R_1>R>R_2$ and $R_2>R$, respectively.

When σ_1 and σ_2 are uniform in the respective regions, $f_{1,n}$ and $f_{2,n}$ are the solutions of differential equations similar to (3.73) in which k is replaced by k_1 and k_2. k_1 and k_2 are defined by

$$k_1^2=4\pi\mu_1\sigma_1 p, \qquad k_2^2=4\pi\mu_2\sigma_2 p. \tag{3.252}$$

The solutions of the differential equations are then given by

$$\begin{aligned} f_{1,n}(\rho)&=\rho^{-1/2}\,[A_n I_\nu(z)+B_n\,K_\nu(z)] \\ f_{2,n}(\rho)&=C_n\,\rho^{-1/2}\,I_\nu(\zeta) \end{aligned} \tag{3.253}$$

where

$$\begin{aligned} \nu&=n+\frac{1}{2} \\ z&=k_1R_1\rho, \quad z_1=k_1R_1, \quad z_q=k_1R_1q \\ \zeta&=k_2R_1\rho, \quad \zeta_q=k_2R_1q. \end{aligned} \tag{3.254}$$

It should be noted that the solution for $f_{2,n}$ is regular at $R=0$. I_ν and K_ν are the modified Bessel functions. For the sake of simplicity, $\mu_1=\mu_2=1$ is assumed in deriving (3.253) and (3.254).

Putting (3.253) into (3.251), we obtain

$$\begin{aligned} &n\,e_n-(n+1)i_n=n(n+1)[A_nI_\nu(z_1)+B_nK_\nu(z_1)] \\ &e_n+i_n=A_n\left[\frac{1}{2}I_\nu(z_1)+z_1I_\nu'(z_1)\right]+B_n\left[\frac{1}{2}K_\nu(z_1)+z_1K_\nu'(z_1)\right] \\ &A_n\left[\frac{1}{2}I_\nu(z_q)+z_qI_\nu'(z_q)\right]+B_n\left[\frac{1}{2}K_\nu(z_q)+z_qK_\nu'(z_q)\right] \\ &\quad=C_n\left[\frac{1}{2}I_\nu(\zeta_q)+\zeta_qI_\nu'(\zeta_q)\right] \\ &A_nI_\nu(z_q)+B_nK_\nu(z_q)=C_nI_\nu(\zeta_q) \end{aligned} \tag{3.255}$$

in which $I_\nu'(z)$ and $K_\nu'(z)$ are respectively $\mathrm{d}I_\nu/\mathrm{d}z$ and $\mathrm{d}K_\nu/\mathrm{d}z$.

Solving the simultaneous equations given in (3.255), we obtain

$$\begin{aligned} \mathrm{A}=&-z_q\{I_{\nu+1}(z_1)K_\nu'(z_q)+K_{\nu+1}(z_1)I_\nu'(z_q)\}I_\nu(\zeta_q) \\ &+\zeta_q\{I_{\nu+1}(z_1)K_\nu(z_q)+K_{\nu+1}(z_1)I_\nu(z_q)\}I_\nu'(\zeta_q) \end{aligned}$$

$$\begin{aligned}
\mathrm{B} &= -z_q\{I_{\nu-1}(z_1)K_\nu'(z_q)+K_{\nu-1}(z_1)I_\nu'(z_q)\}I_\nu(\zeta_q) \\
&\quad +\zeta_q\{I_{\nu-1}(z_1)K_\nu(z_q)+K_{\nu-1}(z_1)I_\nu(z_q)\}I_\nu'(\zeta_q) \\
\frac{i_n}{e_n} &= \frac{n}{n+1}\frac{\mathrm{A}}{\mathrm{B}} \\
\frac{C_n}{e_n} &= \frac{2n+1}{n+1}z_1^{-1}[-z_q\{I_{\nu-1}(z_1)K_\nu'(z_q)+K_{\nu-1}(z_1)I_\nu'(z_q)\}I_\nu(\zeta_q) \\
&\quad +\zeta_q\{I_{\nu-1}(z_1)K_\nu(z_q)+K_{\nu-1}(z_1)I_\nu(z_q)\}I_\nu'(\zeta_q)]^{-1}
\end{aligned} \tag{3.256}$$

from which we can determine i_n and C_n when e_n is given. A_n and B_n can of course be obtained though no explicit expressions are given here.

When $\sigma_2 \to 0$ or $\zeta_q \to 0$, i_n/e_n in (3.256) reduces to

$$\frac{i_n}{e_n}=\frac{n}{n+1}\frac{I_{\nu+1}(z_1)\,K_{\nu+1}(z_q)-K_{\nu+1}(z_1)\,I_{\nu+1}(z_q)}{I_{\nu-1}(z_1)\,K_{\nu+1}(z_q)-K_{\nu-1}(z_1)\,I_{\nu+1}(z_q)}. \tag{3.257}$$

(3.257) agrees with i_n/e_n obtained from (3.79) and (3.82) by putting $\mu=1$. We therefore confirm that the first equation of (3.256) coincides with that for a hollow spherical shell when the conductivity of the space surrounded by the shell is zero.

When $n=1$, the normal (H_R) and tangential (H_θ) components of the magnetic field at $\rho=q$ are from (3.248), (3.253), (3.254) and (3.256) obtained as

$$\begin{aligned}
H_R &= -2\,q^{-3/2}I_{3/2}(\zeta_q)C_1S_1 \\
H_\theta &= q^{-3/2}\Big[I_{3/2}(\zeta_q)-\zeta_q I_{1/2}(\zeta_q)\Big]C_1\frac{\partial S_1}{\partial\theta}
\end{aligned} \tag{3.258}$$

where

$$\begin{aligned}
C_1 &= \frac{3}{2}z_1^{-1}\,\Phi^{-1}\,e_1 \\
\Phi &= -z_q[I_{1/2}(z_1)K_{3/2}'(z_q)+K_{1/2}(z_1)I_{3/2}'(z_q)]\,I_{3/2}(\zeta_q) \\
&\quad +\zeta_q[I_{1/2}(z_1)K_{3/2}(z_q)+K_{1/2}(z_1)I_{3/2}(z_q)]\,I_{3/2}'(\zeta_q).
\end{aligned} \tag{3.259}$$

From Section 3.3 we deduce that the normal ($H_{R,0}$) and tangential ($H_{\theta,0}$) components of the magnetic field at $\rho=q$ become

$$\begin{aligned}
H_{R,0} &= -3\,q^{-3/2}\Psi^{-1}\,e_1S_1 \\
H_{\theta,0} &= -3\,q^{-3/2}\,\Psi^{-1}\,e_1\frac{\partial S_1}{\partial\theta}
\end{aligned} \tag{3.260}$$

when the conductivity for $R_2>R$ is zero. Ψ is given by (3.92).

We thus obtain the ratios of the field components as

$$R_R = H_R/H_{R,0} = z_1^{-1}\, I_{3/2}(z_q)\,\Psi/\Phi$$

and (3.261)

$$R_\theta = H_\theta/H_{\theta,0} = \frac{1}{2} z_1^{-1}[\zeta_q\, I_{1/2}(\zeta_q) - I_{3/2}(\zeta_q)]\,\Psi/\Phi$$

from which we see that the modification for the normal component caused by the presence of the inner conductor is different from that for the tangential component. The situation is much the same as that for a thin spherical sheet underlain by a concerntric perfect conductor at a certain depth studied in Subsection 3.8.1.

Figure 3.36, 3.37, 3.38 and 3.39 show mod R_θ' and mod R_R' for τ $(=\Omega_2/\omega_1)=20$, 10, 5 and 0, respectively. $q=0.8$ is assumed for all the models. Induction parameters Ω_1 and Ω_2 are as usual defined by

$$\Omega_1 = \sqrt{4\pi\sigma_1\omega}\, R_1$$

and

$$\Omega_2 = \sqrt{4\pi\sigma_2\omega}\, R_1$$

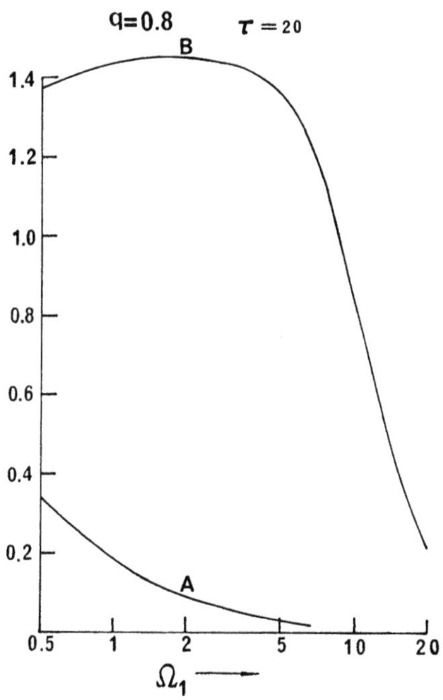

FIG. 3.36 Changes in mod R_θ' and mod R_R' with Ω_1 as indicated by curves A and B respectively. $\tau=20$ and $q=0.8$ are assumed. The physical meaning of R_θ' and R_R' is given in the text.

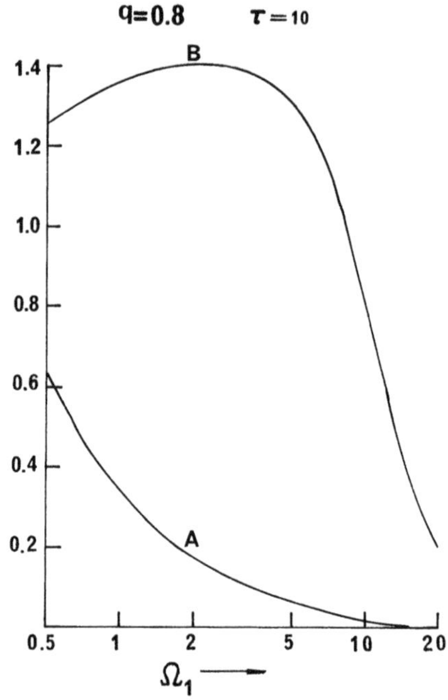

FIG. 3.37 Changes in mod R_θ' and mod R_R' with Ω_1 as indicated by curves A and B respectively. $\tau=10$ and $q=0.8$ are assumed. The physical meaning of R_θ' and R_R' is given in the text.

in which ω is the angular frequency of the oscillating magnetic field.

R_R' and R_θ' are defined by

$$\begin{aligned} R_R' &= 3\, q^{-3/2}\, \Psi^{-1}\, R_R \\ R_\theta' &= 3\, q^{-3/2}\, \Psi^{-1}\, R_\theta \end{aligned} \tag{3.262}$$

so that R_R' and R_θ' represent R_R and R_θ normalized to their respective values at $R=R_1$ when there is no internal conductor.

Looking at these figures, we observe that mod R_θ' takes on a maximum value around 1.4 for large values of τ. It is therefore seen that the magnetic field becomes fairly large when ω is close to $\pi/2$. Actually, the total field is amplified by a factor 1.5 or so for $R_1>R>R_2$. The situation is much the same as that for a thin spherical shell underlain by a concentric perfect conductor as studied in Subsection 3.8.1. When $\theta=0$, the magnetic field becomes very much smaller for large Ω_1 because of the cancelling effect of the inner conductor.

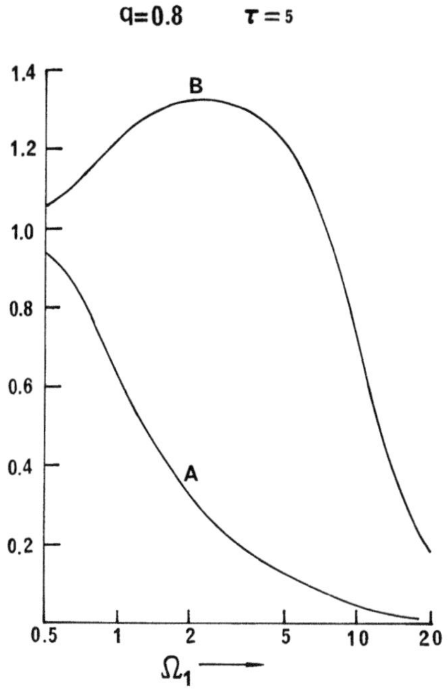

FIG. 3.38 Changes in mod R_θ' and mod R_R' with Ω_1 as indicated by curves A and B respectively. $\tau=5$ and $q=0.8$ are assumed. The physical meaning of R_θ' and R_R' is given in the text.

As it is tedious to present curves for mod R_θ' and mod R_R' for many combinations of parameters q and τ, moduli and arguments for various q and τ obtained as computer output are listed in tables in Appendix 3.

3.8.4 Cylindrical layer model

Electromagnetic shielding by a two-layer cylindrical model of infinite length can be formulated in a manner similar to that for a two-layer spherical model as given in the last subsection. The theory is more or less the same as that developed for a cylindrical shell studied in Subsection 3.4.1.

Let us take a two-layer model for which the cross-section is characterized by outer and inner radii, r_1 and r_2, respectively. Taking a coordinate system (r, θ, z) in which the z axis coincides with the central axis of the cylinder, we can define vector potentials for regions $r_1>r>r_2$ and $r_2>r$. The only non-vanishing components of the vector potentials

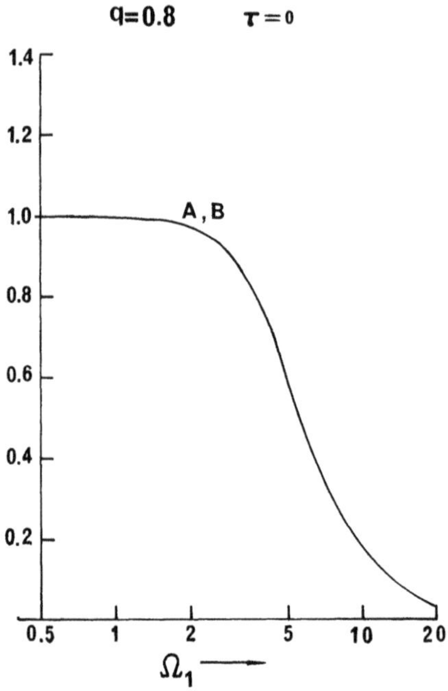

FIG. 3.39 Changes in mod R_θ' and mod R_R' with Ω_1 as indicated by curves A and B respectively. $\tau=0$ and $q=0.8$ are assumed. The physical meaning of R_θ' and R_R' is given in the text. This case corresponds to a hollow shell, so that the curve in the figure coincides with that for mod $\bar{e}_1/e_1$ in Fig. 3.7.

are denoted by $A_{z,1}$ and $A_{z,2}$, respectively for $r_1>r>r_2$ and $r_2>r$. $A_{z,1}$ and $A_{z,2}$ satisfy differential equations similar to (3.111).

Putting

$$\partial/\partial t=p, \quad k_1^{\,2}=4\pi\mu_1\sigma_1 p, \quad k_2^{\,2}=4\pi\mu_2\sigma_2 p$$

typical solutions are obtained as

$$A_{z,1,n}=[D_n I_n(k_1 r)+F_n K_n(k_1 r)] \sin n\theta \quad (r_1>r>r_2) \tag{3.263}$$

and

$$A_{z,2,n}=G_n I_n(k_2 r) \sin n\theta \quad (r_2>r) \tag{3.264}$$

where I_n and K_n are modified Bessel functions. Solutions do not include K_n in (3.264) because the potential should not become infinite at $r=0$. In deriving the above solutions, the electrical conductivities and magnetic permeabilities are assumed to be σ_1 and μ_1 for $r_1>r>r_2$ and σ_2

and μ_2 for $r_2>r$. Only the terms proportional to sin $n\theta$ are taken here. The following also holds good for terms proportional to cos $n\theta$.

From (3.263) and (3.264), the components of magnetic flux density are obtained as

$$\begin{aligned} B_{1,r}&=nr^{-1}[D_nI_n(k_1r)+F_nK_n(k_1r)]\cos n\theta \\ B_{1,\theta}&=-k_1[D_nI_n'(k_1r)+F_nK_n'(k_1r)]\sin n\theta \end{aligned} \qquad (r_1>r>r_2) \quad (3.265)$$

and

$$\begin{aligned} B_{2,r}&=n\,r^{-1}\,G_nI_n(k_2r)\cos n\theta \\ B_{2,\theta}&=-k_2G_nI_n'(k_2r)\sin n\theta \end{aligned} \qquad (r_2>r) \quad (3.266)$$

where $I_n'(z)=\mathrm{d}I/\mathrm{d}z$ and $K_n'(z)=\mathrm{d}K/\mathrm{d}z$.

In the non-conducting region $r>r_1$, we can define a magnetic potential W of which a typical term is given by

$$W_{0,n}=(a_nr^n+A_nr^{-n})\cos n\theta \qquad (r>r_1) \quad (3.267)$$

which satisfies $\nabla^2W=0$. The magnetic field components are then obtained as

$$\begin{aligned} H_{0,r}&=-n\,r^{-1}(a_nr^n-A_nr^{-n})\cos n\theta \\ H_{0,\theta}&=nr^{-1}(a_nr^n+A_nr^{-n})\sin n\theta \end{aligned} \qquad (r>r_1). \quad (3.268)$$

The continuity conditions for magnetic field and magnetic flux density at $r=r_1$ and r_2 imply

$$\begin{aligned} &a_nr_1^n-A_nr_1^{-n}=-[D_nI_n(k_1r_1)+F_nK_n(k_1r_1)] \\ &\mu n(a_nr_1^n+A_nr_1^{-n})=-k_1r_1[D_nI_n'(k_1r_1)+F_nK_n'(k_1r_1)] \\ &D_nI_n(k_1r_2)+F_nK_n(k_1r_2)=G_nI_n(k_2r_2) \\ &k_1[D_nI_n'(k_1r_2)+F_nK_n'(k_1r_2)]=\mu\,k_2G_nI_n'(k_2r_2) \end{aligned} \qquad (3.269)$$

where $\mu_1=\mu_2=\mu$ is assumed.

Solving (3.269), we obtain

$$\begin{aligned} 2\,n\mu r_1^n\frac{a_n}{G_n}&=[z_1I_n'(z_1)+n\,\mu I_n(z_1)][\mu\,\zeta_qK_n(z_q)I_n'(\zeta_q)-z_qK_n'(z_q)\,I_n(\zeta_q)] \\ &-[z_1K_n'(z_1)+n\mu\,K_n(z_1)][\mu\,\zeta_qI_n(z_q)I_n'(\zeta_q)-z_qI_n'(z_q)I_n(\zeta_q)] \end{aligned} \qquad (3.270)$$

where

$$z_1=k_1r_1,\quad z_q=k_1r_2,\quad \zeta_q=k_2r_2,\quad q=r_2/r_1. \qquad (3.271)$$

When $n=1$ and $\mu=1$, (3.270) reduces to

$$\frac{G_1}{a_1}=-2\,r_1z_1^{-1}\,\Phi_c^{-1} \qquad (3.272)$$

where

$$\Phi_c = -z_q I_1(\zeta_q)[I_0(z_1)K_1'(z_q)+K_0(z_1)I_1'(z_q)] \\ +\zeta_q I_1'(\zeta_q)[I_0(z_1)K_1(z_q)+K_0(z_1)I_1(z_q)]. \tag{3.273}$$

We also obtain

$$\frac{A_1}{r_1^2 a_1} = [\zeta_q I_1'(\zeta_q)\{I_2(z_1)K_1(z_q)+K_2(z_1)I_1(z_q)\} \\ -z_q I_1(\zeta_q)\{I_2(z_1)K_1'(z_q)+K_2(z_1)I_1'(z_q)\}]\Phi_c^{-1} \tag{3.274}$$

which is useful for estimating the electromagnetic response of the system.

The normal (H_r) and tangential (H_θ) components of the magnetic field at $r=r_2$ for $n=1$ and $\mu=1$ are from (3.266), (3.270) and (3.271) obtained as

$$H_r = r_2^{-1} G_1 I_1(\zeta_q) a_1 \cos\theta \\ H_\theta = -r_2^{-1} G_1 \zeta_q I_1'(\zeta_q) a_1 \sin\theta \tag{3.275}$$

where G_1 is obtained from (3.272) and (3.273).

When the region defined by $r_2>r$ is non-conducting, the normal ($H_{r,0}$) and tangential ($H_{\theta,0}$) components of the magnetic field at $r=r_2$ can be obtained as in Section 3.4 as

$$H_{r,0} = -2\,q^{-1}\Psi_c^{-1}a_1\cos\theta$$

and

$$H_{\theta,0} = 2\,q^{-1}\,\Psi_c^{-1}a_1\sin\theta \tag{3.276}$$

where Ψ_c is given by (3.134).

We thus obtain the ratios of the magnetic field components as

$$R_r = H_r/H_{r,0} = z_1^{-1} I_1(\zeta_q)\Psi_c/\Phi_c \\ R_\theta = H_\theta/H_{\theta,0} = z_1^{-1}[\zeta_q I_0(\zeta_q) - I_1(\zeta_q)]\Psi_c/\Phi_c\,. \tag{3.277}$$

As noted in the last subsection with regard to spherical models, the influence of the inner conductor on the r component of the magnetic field is different from that on the θ component.

We shall define R_r' and R_θ' which are normalized by their respective values at $r=r_1$ when there is no inner conductor. In that case we have

$$R_r' = 2\,q^{-1}\Psi_c^{-1}R_r \\ R_\theta' = 2q^{-1}\Psi_c^{-1}R_\theta\,. \tag{3.278}$$

On defining induction parameters Ω_1 and Ω_2 in a way similar to that in the last subsection just replacing R by r, R_r' and R_θ' are readily calculated for various combinations of q and τ ($=\Omega_2/\Omega_1$). Since it is much too tedious to present R_r' and R_θ' numerically and graphically, it is only pointed out here that the mode of dependence of their moduli and arguments on various combinations of q and τ is more or less similar to that for the spherical layer models as given in the tables in Appendix 3. If one wants to determine rigorous values of R_r', R_θ' and $A_1/r_1^2a_1$, the equations cited in this subsections may be of use.

Generally speaking, the behaviour of magnetic lines of force for a cylindrical two-layer model is more or less the same as that for a spherical two-layer model. What was stated in the last subsection therefore holds in this case to a good approximation.

3.8.5 Negi's "negative screening"

Let us go back to Subsection 3.8.3 and examine how the electromagnetic response mod i_n/e_n of spherical two-layer models as given in (3.256) changes with the induction parameter. On the basis of the computer output listed in the tables in Appendix 3, mod i_1/e_1 for various values of Ω_1 is plotted in Fig. 3.40 for q=0.8 and 0.4 taking τ as a parameter.

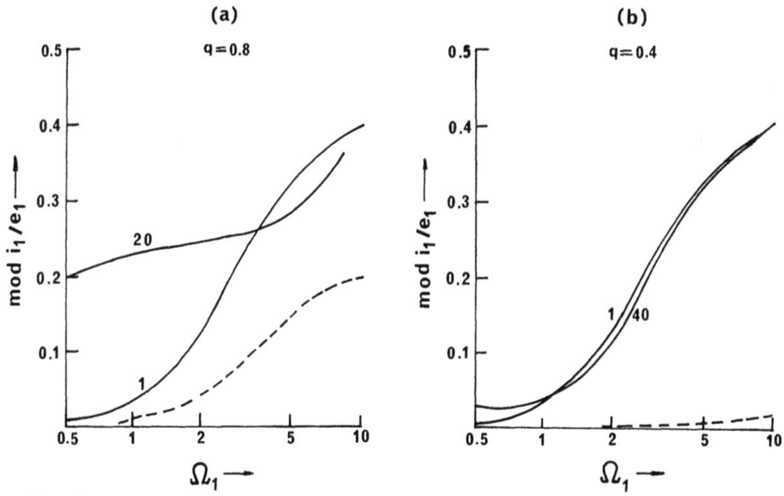

FIG. 3.40 Changes in electromagnetic response with induction parameter Ω_1 for two-layer spherical shells. Two cases for which q=0.8 and 0.4 are sudied. The numbers attached to solid curves indicate values of τ ($=\Omega_2/\Omega_1$). Dashed curves indicate the responses at $R=R_1$, where $q=1$, when the outer layer is non-conducting.

We observe in Fig. 3.40 (b) that the response for $\tau=40$ is almost the same as that for $\tau=1$ (uniform sphere) when Ω_1 is larger than 1. The radius of the inner sphere of high conductivity is small in this case, so that the influence of the inner conductor is negligible for a high-frequency oscillating magnetic field.

In contrast to Fig. 3.40 (b), the effect of the inner sphere of high conductivity becomes apparent when the radius of the inner conductor becomes larger as can be seen in Fig. 3.40 (a) for which we assume $q=0.8$. The response becomes quite large for a smaller induction parameter in this case. This is quite natural because of the deeper penetration of induced electric currents. However, we see that the response for $\tau=20$ when Ω_1 is larger than 3 or so is smaller than that for $\tau=1$. Such a phenomenon is probably caused by the fact that much of the induced currents flows in the inner conductor. In that case the distance between the effective depth at which the currents are flowing and the outer surface of the superficial layer becomes large, and so the magnetic field produced by these currents becomes small at the surface.

When the outer layer defined by $R_1>R>R_2$ is non-conducting, the response of the inner sphere, that occupies the region $R_2>R$, are readily obtained at $R=R_1$ as shown by the dashed lines in Fig. 3.40 (a) and 3.40 (b) for $q=0.8$ and 0.4, respectively. We see in the figure that the fact that the conductor lies at a large depth makes the response very small.

NEGI (1967), NEGI and GUPTA (1968), NEGI and RAVAL (1969), NEGI and VERMA (1972) and GUPTA *et al.* (1973) studied electromagnetic induction in spherical and cylindrical two-layer models. NEGI (1967) represented the response of the inner spherical conductor in the presence of an overburden or an outer layer by

$$C_{1,R_2}=C_1-C_{1,R_1} \tag{3.279}$$

where the notations involved are defined by:

C_1: The response of the two-layer system as a whole;

C_{1,R_1}: The contribution to the response from the homogeneous sphere of radius R_1;

C_{1,R_2}: The contribution to the response from the inhomogeneity present in the large sphere of radius R_1, in the form of the sphere of radius R_2.

It is assumed here that the inducing field is uniform.

In order to study the influence of the outer layer on the in-phase component, Negi defined δ_{Re} by

$$\delta_{\mathrm{Re}} = -\frac{\mathrm{Re}C_{1,R_2}}{\mathrm{Re}C_{0,R_2}} \tag{3.280}$$

where C_{0,R_2} corresponds to the response when the outer layer is replaced by a non-conductor. The negative sign has been included to indicate the screening behaviour of the outer layer.

Based on the computer output given in Appendix 3, δ_{Re} for $q=0.4$ is calculated as a function of the induction parameter and is shown in Fig. 3.41. Although this is only an example, we see that δ_{Re} exceeds unity. This would mean, as NEGI (1967) pointed out, that the inner body behaves as if it is almost doubly conducting in the presence of the overburden or outer layer.

It is therefore seen that the outer layer does not shield the inner sphere. On the contrary, the outer layer apparently enhances the electromagnetic response of the inner conductor. NEGI (1967) argued that an enhancement of response such as this is quite a curious phenomenon, and named this phenomenon a kind of negative screening.

It is the author's opinion, however, that the phenomenon pointed out by Negi arises naturally judging from the behaviour of mod i_1/e_1 for larger values of τ as shown in Fig. 3.40.

Attention should be drawn to the wording "negative screening". This might be taken to suggest an enhancement of the magnetic field

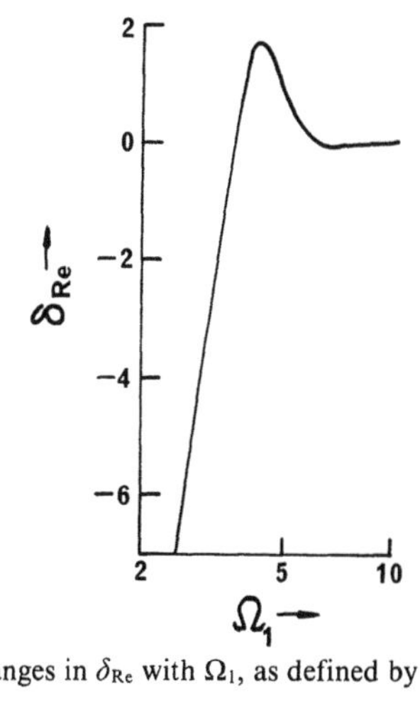

FIG. 3.41 Changes in δ_{Re} with Ω_1, as defined by (3.280), for $q=0.4$.

immediately inside the outer conducting layer, to an anomalous extent. But this is not so. The rate of screening has been defined by Eq. (3.280). Contrary to Negi, the author thinks that nothing curious is associated with the problem of electromagnetic induction in a two-layer conductor.

Chapter 4

ELECTROMAGNETIC INDUCTION IN A NON-UNIFORM SHEET

No general theory of electromagnetic induction in a conductor having a heterogeneously-distributed conductivity has so far been developed because of mathematical difficulties due to the fact that electric and magnetic fields cannot be treated separately from charge distributions. When the conductor can be treated as a thin sheet or when its thickness is much smaller than the wavelength of the inducing magnetic field, such a difficulty can be avoided, so that a general theory of electromagnetic induction in a non-uniform thin sheet can be advanced.

Electromagnetic shielding by a non-uniform thin sheet is of little importance for practical applications of shielding because approximately uniform material can always be used for that purpose. In spite of that fact, it is of some interest from a theoretical standpoint to investigate electromagnetic induction in a non-uniform sheet. The theory of electromagnetic induction in a non-uniform sheet was developed by PRICE (1949), ASHOUR and PRICE (1948) and others. Some supplements of the theory were presented by RIKITAKE (1966).

One of the most interesting and rather unusual features with regard to electromagnetic induction in a non-uniform sheet is certainly a resonance-like enhancement of the induced electric currents for a particular combination of the distribution of conductivity in the sheet and the wavelength of the inducing magneitc field, as pointed out by RIKITAKE (1968) who studied electormagnetic induction in a sheet in which the conductivity changes periodically in one direction. A theory of electromagnetic induction in a sheet underlain by a semi-infinite conductor as shown in Fig. 4.1 has been developed by him.

Let us assume an inducing field specified by the magnetic potential

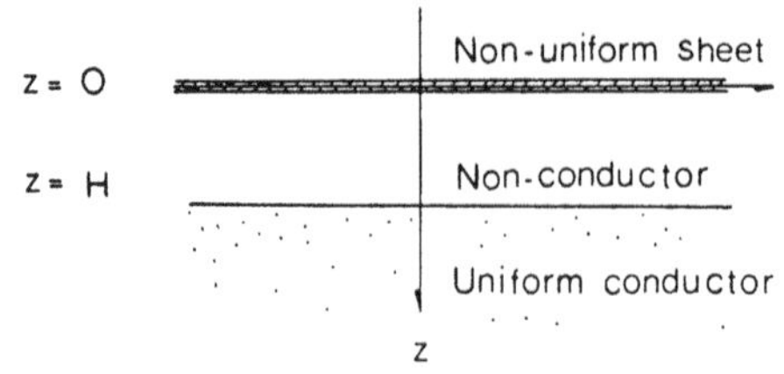

FIG. 4.1 System of conductors.

$$W_e = A\mathrm{e}^{-qz}\cos qy. \tag{4.1}$$

When the distribution of electrical conductivity in the sheet is symmetric about the origin, it is obvious that only an induced field having a potential expressed by a sum of harmonic terms proportional to cosmpx results. The magnetic potentials in regions $z<0$ and $0<z<H$ are therefore written as

$$W_1 = \{A\mathrm{e}^{-qz} + \sum_m \exp\,[(m^2p^2+q^2)^{1/2}z]i_m \cdot \cos mpx\} \cos qy \qquad (z<0), \tag{4.2}$$

and

$$W_2 = \sum_m \{\exp\,[-m^2p^2+q^2)^{1/2}z]e_m' + \exp\,[(m^2p^2+q^2)^{1/2}z]i_m'\} \cdot \cos mpx \cos qy \qquad (0<z<H). \tag{4.3}$$

In the conducting region for $H<z$ (conductivity$=\sigma$), we may define a vector potential $\boldsymbol{A}$ that satisfies $\nabla^2\boldsymbol{A}=4\pi\sigma\partial\boldsymbol{A}/\partial t$. The x, y and z ocmponents of a typical solution of $\boldsymbol{A}$ are given as

$$\boldsymbol{A}_m = \begin{cases} qC_m \exp\,[-(m^2p^2+q^2+\chi^2)^{1/2}z] \cos mpx \sin qy \\ -mpC_m \exp\,[-(m^2p^2+q^2+\chi^2)^{1/2}z] \sin mpx \cos qy \\ 0 \end{cases} \qquad (H<z) \tag{4.4}$$

where putting $\partial/\partial t = D$, χ is defined by

$$\chi^2 = 4\pi\sigma D. \tag{4.5}$$

The author should like to draw reader's attention to the point that notations D, p, q, μ and the like here have different meanings from those used in the previous chapters. But these notations are clearly defined in this chapter.

In the above derivation, the magnetic permeability is assumed as unity in electromagnetic units. We further assume that induced currents, expressed by a current function Ψ, are flowing in the sheet. Ψ is generally given as

$$\Psi=\sum_m K_m \cos mpx \cos qy. \tag{4.6}$$

By differentiating the potentials in the x, y and z directions and taking the relation curl $\boldsymbol{A}=\boldsymbol{H}$ into account, the three components of the typical magnetic field in the respective regions are obtained as

$$\boldsymbol{H}_1=\begin{cases} mp \exp\,[(m^2p^2+q^2)^{1/2}z]\, i_m \sin mpx \cos qy \\ q\{\lambda_m A e^{-qz}+\exp\,[(m^2p^2+q^2)^{1/2}z]\, i_m \cos mpx\} \sin qy \qquad (z<0) \\ \{\lambda_m q A e^{-qz}-(m^2p^2+q^2)^{1/2} \\ \quad \cdot \exp\,[(m^2p^2+q^2)^{1/2}z]\, i_m \cos mpx\} \cos qy \end{cases} \tag{4.7}$$

$$\boldsymbol{H}_2=\begin{cases} mp\{\exp[-(m^2p^2+q^2)^{1/2}z]e_m' \\ \quad +\exp\,[(m^2p^2+q^2)^{1/2}z]\, i_m'\} \sin mpx \cos qy \\ q\{\exp\,[-(m^2p^2+q^2)^{1/2}z]e_m' \qquad (0<z<H) \\ \quad +\exp\,[(m^2p^2+q^2)^{1/2}z]i_m'\} \cos mpx \sin qy \\ (m^2p^2+q^2)^{1/2}\{\exp\,[-(m^2p^2+q^2)^{1/2}z]\, e_m' \\ \quad -\exp\,[(m^2p^2+q^2)^{1/2}z]\, i_m'\} \cos mpx \cos qy \end{cases} \tag{4.8}$$

$$\boldsymbol{H}_3=\begin{cases} mp(m^2p^2+q^2+\chi^2)^{1/2}\, C_m \\ \quad \cdot \exp\,[-(m^2p^2+q^2+\chi^2)^{1/2}z] \sin mpx \cos qy \\ q(m^2p^2+q^2+\chi^2)^{1/2}\, C_m \qquad (H<z) \\ \quad \cdot \exp\,[-(m^2p^2+q^2+\chi^2)^{1/2}z] \cos mpx \sin qy \\ (m^2p^2+q^2)\, C_m \\ \quad \cdot \exp\,[-(m^2p^2+q^2+\chi^2)^{1/2}z] \cos mpx \cos qy \end{cases} \tag{4.9}$$

where

$$\begin{aligned} \lambda_m&=1 \qquad m=0 \\ \lambda_m&=0 \qquad m\neq 0. \end{aligned} \tag{4.10}$$

The continuity conditions of the magnetic fields at $z=0$ and $z=H$ imply

$$\begin{aligned}
&\lambda_m A + i_m = e_m' + i_m' + 4\pi K_m \\
&\lambda_m A - i_m = e_m' - i_m' \\
&\exp\,[-(m^2p^2+q^2)^{1/2}H]\; e_m' + \exp\,[(m^2p^2+q^2)^{1/2}H]\; i_m' \\
&\quad = (m^2p^2+q^2+\chi^2)^{1/2} \exp\,[-(m^2p^2+q^2+\chi^2)^{1/2}\,H]\; C_m \\
&\exp\,[-(m^2p^2+q^2)^{1/2}H]\; e_m' - \exp\,[(m^2p^2+q^2)^{1/2}H]\; i_m' \\
&\quad = (m^2p^2+q^2)^{1/2} \exp\,[-(m^2p^2+q^2+\chi^2)^{1/2}\,H]\; C_m .
\end{aligned} \tag{4.11}$$

The condition that should be satisfied by Ψ is known to be

$$\begin{aligned}
&\rho(\partial^2\Psi/\partial x^2 + \partial^2\Psi/\partial y^2) + (\mathrm{d}\rho/\mathrm{d}x)\; \partial\Psi/\partial x \\
&\quad = \partial^2 W_1/\partial z\partial t \qquad\qquad (z=0)
\end{aligned} \tag{4.12}$$

in which $\rho=\rho\,(x)$ denotes the resistance of the sheet. Putting (4.2) and (4.6) into (4.12), we obtain

$$\begin{aligned}
&\sum_m [\rho(m^2p^2+q^2)\cos mpx \\
&\quad + (\mathrm{d}\rho/\mathrm{d}x)\; mp \sin mpx]\; K_m \\
&\quad = D\sum_m (m^2p^2+q^2)^{1/2}(\lambda_m A - i_m) \cos mpx .
\end{aligned} \tag{4.13}$$

It is possible to derive the following relation from the equations in (4.11);

$$\lambda_m A - i_m = (\lambda_m A - 2\pi K_m)\Phi_m \tag{4.14}$$

where

$$\begin{aligned}
\Phi_m = 1 - &\frac{(m^2p^2+q^2+\chi^2)^{1/2} - (m^2p^2+q^2)^{1/2}}{(m^2p^2+q^2+\chi^2)^{1/2} + (m^2p^2+q^2)^{1/2}} \\
&\cdot \exp[-2(m^2p^2+q^2)^{1/2}H].
\end{aligned} \tag{4.15}$$

Introducing (4.14) into (4.13), we obtain

$$\begin{aligned}
&\sum_m [\rho(m^2p^2+q^2)\cos mpx \\
&\quad + (\mathrm{d}\rho/\mathrm{d}x)\; mp \sin mpx] K_m \\
&\quad = D\sum_m (m^2p^2+q^2)^{1/2}(\lambda_m A - 2\pi K_m)\Phi_m \\
&\quad \cdot \cos mpx
\end{aligned} \tag{4.16}$$

from which we may obtain K_m.

When $\sigma \to 0$ $(\chi \to 0)$ or $H \to \infty$, we see that Φ_m tends to 1, so that (4.16) reduces to the result for the same sheet placed in free space.

The resistance in the sheet is assumed to be expressed as

$$\rho = \rho_0(1 - \epsilon \cos rx) \qquad \epsilon < 1 \tag{4.17}$$

in which ρ_0 is a constant resistance.

Equation (4.16) into which Eq. (4.17) is substituted is multiplied by $\cos Mpx$ and integrated with respect to x from $-\pi/p$ to π/p. In that case we obtain

$$\begin{aligned} &\rho_0\pi(M^2p^2+q^2)\,K_M \\ &-\rho_0\epsilon \sum_m [(m^2p^2+q^2)\,I_1 - mp^2\nu I_2]\,K_m \\ &= D\pi(\lambda_M A - 2\pi K_M)(M^2p^2+q^2)^{1/2}\Phi_M \end{aligned} \tag{4.18}$$

where

$$\begin{aligned} &\nu = r/p \\ &I_1 = \int_{-\pi}^{\pi} \cos \nu s \cos ms \cos Ms \, ds \\ &I_2 = \int_{-\pi}^{\pi} \sin \nu s \sin ms \cos Ms \, ds\,. \end{aligned} \tag{4.19}$$

The integrals involved in (4.19) are readily calculated as follows;

$$\begin{aligned} I_1 &= J_1 + J_2 + J_3 + J_4 \\ I_2 &= J_2 - J_1 - J_3 + J_4 \end{aligned} \tag{4.20}$$

in which

$$J_1 = \begin{cases} \dfrac{1}{2}\dfrac{1}{\nu+m+M}\sin(\nu+m+M)\pi & \nu+m+M \neq 0 \\ \pi/2 & \nu+m+M = 0 \end{cases}$$

$$J_2 = \begin{cases} \dfrac{1}{2}\dfrac{1}{\nu-m-M}\sin(\nu-m-M)\pi & \nu-m-M \neq 0 \\ \pi/2 & \nu-m-M = 0 \end{cases}$$

$$J_3 = \begin{cases} \dfrac{1}{2}\dfrac{1}{\nu+m-M}\sin(\nu+m-M)\pi & \nu+m-M \neq 0 \\ \pi/2 & \nu+m-M = 0 \end{cases}$$

$$J_4=\begin{cases}\dfrac{1}{2}\dfrac{1}{\nu-m+M}\sin(\nu-m+M)\pi & \nu-m+M\neq 0\\ \pi/2 & \nu-m+M=0.\end{cases} \tag{4.21}$$

If we ignore terms for large values of m and M, Eq. (4.18) provides a set of simultaneous equations from which we may determine K_0, K_1, K_2, ··· for a given ν.

When the magnetic variation is purely periodic in time, we put

$$D=i\alpha \tag{4.22}$$

in which

$$\alpha=2\pi/T \tag{4.23}$$

T being period.

When we put

$$\begin{aligned}K_m&=\bar{K}_m+iK_m{}^*\\ \Phi_m&=\bar{\Phi}_m+i\Phi_m{}^*,\end{aligned} \tag{4.24}$$

(4.18) separates into the following two equations;

$$\begin{aligned}&\epsilon\sum_m[(m^2+\mu^2)I_1-m\nu I_2]\bar{K}_m\\ &=\pi(M^2+\mu^2)\bar{K}_M+\beta(M^2+\mu^2)^{1/2}\\ &\cdot(\lambda_M A\Phi_M{}^*-2\pi\Phi_M{}^*\bar{K}_M-2\pi\bar{\Phi}_M K_M{}^*)\end{aligned} \tag{4.25a}$$

$$\begin{aligned}&\epsilon\sum_m[(m^2+\mu^2)I_1-m\nu I_2]K_m{}^*\\ &=\pi(M^2+\mu^2)K_M{}^*-\beta(M^2+\mu^2)^{1/2}\\ &\cdot(\lambda_M A\bar{\Phi}_M-2\pi\bar{\Phi}_M\bar{K}_M+2\pi\Phi_M{}^*K_M{}^*)\end{aligned} \tag{4.25b}$$

where

$$\beta=\pi\alpha/\rho_0 p,\qquad \mu=q/p. \tag{4.26}$$

Real and imaginary current function coefficients for various values of ν, H, and T are calculated. It is assumed that the underlying half-space is perfectly conducting, that $\epsilon=0.99$, which gives a maximum conductivity ocntrast of 199:1 in the sheet, that ρ_0 amounts to 10^{-2} ohm m ($=10^8$ e. m. u.) and that $1/p=0.5577\times10^8$ cm. The last assumption corresponds to a wavelength of 40° in longitude, provided that the induction takes place at midlatitude of the earth (38° in latitude, say). The coefficients for $m>23$ are ignored in the actual calculation, so that simultaneous equations with forty-eight unknowns are solved. Throughout the calculation, $\mu=1$ is assumed.

The present theory assumes that the resistance is periodic in the x direction with a period of $2\pi/p$. The conductivity distributions in the sheet for a number of different values of ν are shown in Fig. 4.2. We see that, for distributions of ν between 0 and 1, both the conductivity contrast and the mean resistance increase as ν gets larger although the fundamental period of the distribution is always the same. For ν greater than 1, the values of ν indicate the ratio of the wavelength of the inducing field to the wavelength of the distribution of resistance, because $\mu=1$ is assumed. All the existing theories of electromagnetic induction in a non-uniform plane sheets (RIKITAKE, 1966; PRICE, 1949) were presented only for $\nu=1$ and $\mu=1$.

Figure 4.3 shows the changes with ν in the imaginary and real coefficients for $m=0$ and $T=600$ sec, as calculated for different values of H, the distance between the sheet and the underlying perfect conductor. It is interesting to note that $\bar{K}_0$ takes on maximum values around $\nu=0.7$–0.8, whereas $K_0{}^*$ changes sign there. It is not entirely clear why we observe such a resonance-like enhancement of the induced current for that particular value of ν.

Changes with ν in the coefficients for $m=0$ are illustrated in the figure because thery are generally larger than the coefficients for $m>0$. It may be said that K_0 provides a good approximation to the actual current function Ψ at least for rapid variations. It is obvious, however, that local behaviour of the induced currents cannot be represented by

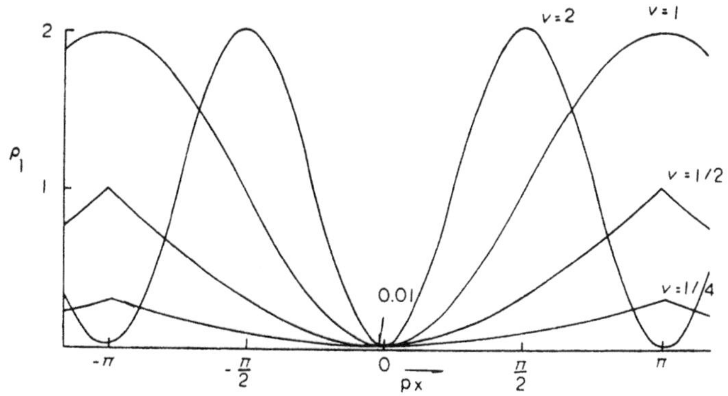

FIG. 4.2 Distributions of dimensionless resistance in the sheet for a few values of ν.

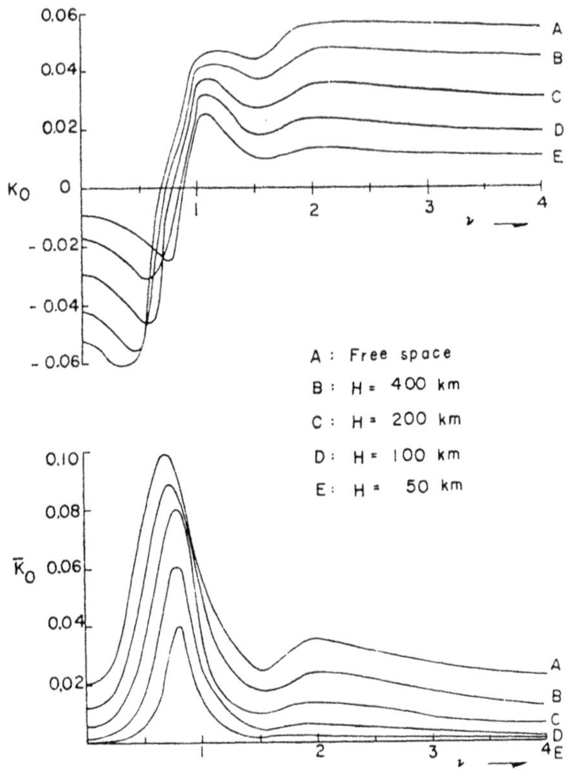

FIG. 4.3 Imaginary and real parts of current function coefficients for $m=0$ and $T=600$ sec, as calculated for different values of H.

K_0 only. The extent of the resonance would certainly depend on the wavelength of the induced currents, so that the resonant behaviour should be examined for each value of m, although it is beyond the scope of this chapter to look into such local behaviour.

The mean resistance of the sheet for $\nu=0.75$ is about 70 times as large as the resistance for $\nu=0$. Nevertheless, the intensity of the induced current for $\nu=0.75$ is far larger than that for $\nu=0$. Looking at the resistnace distributions for various values of ν less than 1, we surmise that the contrast between the lowest and highest values of resistance largely controls the resonant behaviour.

The influence of the underlying conductor on the induced currents in the sheet can be seen from Fig. 4.3, too. It is observed in the figure that the damping effect due to the underlying conductor prevails except for ν close to the value for the resonance. Such a damping effect has been generally known (e.g., RIKITAKE, 1961), the effect usually becoming large as the conductor approaches the sheet. As can be seen from Fig. 4.3, however, the damping effect around the resonance state seems so anomalous that very little decrease in the current function coefficient is observed for larger values of H. Actually, there is practically no decrease in $\bar{K}_0$ for H=400 km around ν=0.9. A similar anomaly in the damping effect can also be seen in the real coefficient for T=60 sec, and the absolute values of the imaginary coefficient for a sheet placed in free space are smaller than those for H=200 km.

The absolute intensity of the induced current can readily be calculated from $\bar{K}_0$ and K_0^*. Changes with ν in the absolute intensity would be somewhat like the curves for $\bar{K}_0$ as shown in the lower half of Fig. 4.3, although the peaks at the resonance would become somewhat flat.

In the light of the above results, we notice that there are certain combinations of conductor configuration, wavelength, and period of the inducing field for which the usual concept of damping does not hold.

No detailed study on the period for which the resonant response becomes most intense has been made. A preliminary study suggests, however, that the resonance prevails at a period that is close to the relaxation time of free decay of electric currents in the sheet. The decay time in this case is defined to be that of a uniform sheet for which the resistance is equal to the averaged resistance of the non-uniform sheet in question.

The author (RIKITAKE, 1961, 1966) has emphasized that the conducting part of the earth's mantle would influence the electric currents induced in the oceans by geomagnetic variations in such a way as to make the current intensity small. The present study suggests, however, that there may be exceptional cases, depending on distributions of the conductivity and on the inducing field, for which the damping effect is almost non-existent. A study of electromagnetic induction by S_q, the daily variation of the geomagnetic field, in a realistic land-sea model (T. Rikitake and T. Yukutake, 1968, unpublished) suggests that even a negative damping effect is possible for particular cases, although no general argument for the conditions for such an effect has been developed.

The resonance-like response of the induced currents in a sheet may sometimes have an important bearing on interpreting geomagnetic variation anomalies. When an anomalously large magnetic variation field arising from inside the earth is observed, the possibility of accounting for it by the anomalous response of a near-surface conductor, as discussed here, should be carefully considered.

Appendix 1

MAGNETIC SHIELDING BY A PLANE SLAB

An account of magnetic shielding by a highly-permeable plane slab of infinite extent was not given in Chapter 2 because the theory for such a case is relatively simple. It is considered, however, that a brief account of magnetic shielding for such a case should be presented since it provides a useful check on electromagnetic shielding when the frequency of the varying field approaches zero.

Let us take a model similar to that shown in Fig. 3.1 in which the conductivity for the region $0>z>-L$ is zero though the medium is permeable there. In the free spaces $z<-L$ and $0<z$, the permeability is taken as unity. For such a model, we can define magnetic potentials W, W' and $\overline{W}$ for the respective regions. The typical solutions for them are given as

$$\begin{aligned} W_{mn}&=(e_{mn}\mathrm{e}^{-kz}+i_{mn}\mathrm{e}^{kz})\,V_{mn} && -L>z \\ W_{mn}{}'&=(A_{mn}\mathrm{e}^{-kz}+B_{mn}\mathrm{e}^{kz})\,V_{mn} && 0>z>-L \\ \overline{W}_{mn}&=\bar{e}_{mn}\mathrm{e}^{-kz}\,V_{mn} && z>0 \end{aligned} \tag{A1.1}$$

on the condition that there are no sources that produce magnetic fields in the region defined by $z>0$.

V_{mn} is a harmonic function that satisfies Eq. (2.21) and

$$k^2=(m^2+n^2)(2\pi/l)^2 \tag{A1.2}$$

as already given by (3.20). l is the wavelength of the magnetic field.

Continuity conditions of the tangential component of magnetic field and of the normal component of magnetic flux density at $z=-L$ and $z=0$ imply

$$
\begin{aligned}
&e\,e^{kL}+i\,e^{-kL}=A\,e^{kL}+B\,e^{-kL}\\
&-e\,e^{kL}+i\,e^{-kL}=\mu(-A\,e^{kL}+B\,e^{-kL})\\
&A+B=\bar{e}\\
&\mu(-A+B)=-\bar{e}
\end{aligned}
\qquad \text{(A1.3)}
$$

solving which we obtain

$$
\begin{aligned}
&A=\frac{\mu+1}{2\mu}\,\bar{e}\\
&B=\frac{\mu-1}{2\mu}\,\bar{e}\\
&\bar{e}/e=4\mu\,e^{kL}/\Psi\\
&i/e=(1-\mu^2)(e^{kL}-e^{-kL})\,e^{2kL}/\Psi
\end{aligned}
\qquad \text{(A1.4)}
$$

where

$$\Psi=(1+\mu)^2e^{kL}-(1-\mu)^2\,e^{-kL}. \qquad \text{(A1.5)}$$

In the expressions (A1.3) and (A1.4), the suffixes m and n are dropped for the sake of simplicity.

As k is defined by Eq. (3.20), we take a magnetic field of the smallest wave-number. In that case, we put

$$n=1,\ m=0.$$

Changes in $\overline{e}/e$ and $(i/e)e^{-2kL}$ with μ can readily be estimated taking kL as a parameter. Figure A1 shows how $\overline{e}/e$ changes as μ increases. It is clearly seen in the figure that the amplitude ratio $\overline{e}/e$ becomes smaller as μ gets larger. This means that the larger the permeability, the larger the shielding effect of the slab. It is also observed in the figure that the shielding effect is greater for a magnetic field having a shorter wavelength than for that having a longer wavelength.

$(i/e)\ e^{-2kL}$ as shown in Fig. A2 is useful for comparing the field arising from the induced magnetization in the slab to the external inducing field at the upper boundary of the slab, where $z=-L$. It is clearly observed that the induced field becomes larger as the permeability becomes higher. Such an effect is more pronounced for a magnetic field having a shorter wavelength.

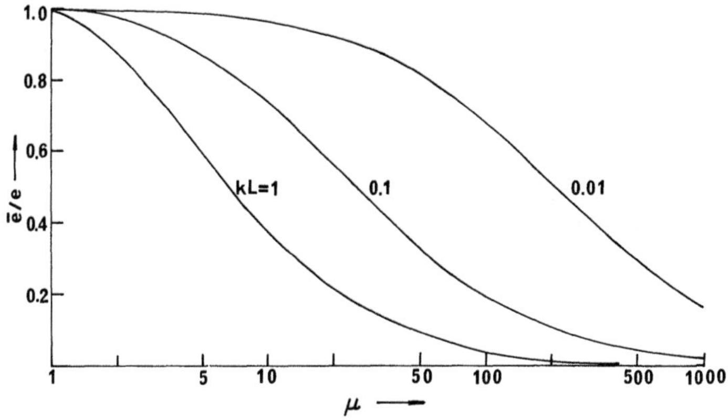

FIG. A1 Changes in $\bar{e}/e$ with μ for a plane slab of magnetic permeability μ. kL is taken as a parameter.

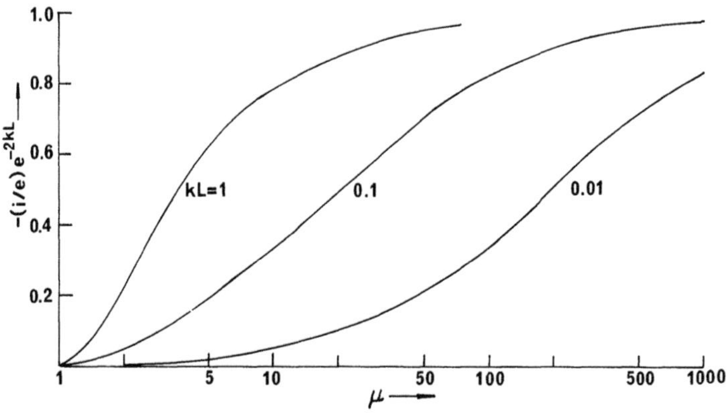

FIG. A2 Changes in $(i/e)\,e^{-2kL}$ with μ for a plane slab of magnetic permeability μ. kL is taken as a parameter.

Appendix 2

ELECTROMAGNETIC SHIELDING BY A PLANE SLAB WHEN IT IS UNDERLAIN BY A CONDUCTOR FROM WHICH THE INDUCING FIELD ARISES

Let us take two-layer model which is the same as that in Fig. 3.34. The following discussion is different from that developed in Subsection 3.8.2 by there being no sources of magnetic field in the region $z<-L$. The inducing field is assumed to arise from the region $z>0$. The magnetic permeability is assumed unity everywhere in electromagnetic units.

In that case, we define a typical magnetic potential such as

$$W_{mn}=i_{mn}\mathrm{e}^{kz}V_{mn} \qquad (z<-L) \tag{A2.1}$$

and so the magnetic field components are given by

$$\boldsymbol{H}=\begin{cases} -i_{mn}\,\mathrm{e}^{kz}\,\dfrac{\partial V_{mn}}{\partial x} \\ -i_{mn}\,\mathrm{e}^{kz}\,\dfrac{\partial V_{mn}}{\partial y} \\ -k\,i_{mn}\mathrm{e}^{kz}\,V_{mn} \end{cases} \qquad (z<-L). \tag{A2.2}$$

In the conducting regions $-L<z<0$ and $0<z$, we define vector potentials $\boldsymbol{A}_1$ and $\boldsymbol{A}_2$ which satisfy the two equations given in (3.230), respectively.

Typical solutions of the two equations are given by

$$\boldsymbol{A}_{1,mn}=(C_{mn}\mathrm{e}^{-\sqrt{k^2+\chi_1{}^2}\,z}+D_{mn}\mathrm{e}^{\sqrt{k^2+\chi_1{}^2}\,z})\cdot(\boldsymbol{r}\wedge\mathrm{grad}\ V_{mn}) \qquad (-L<z<0) \tag{A2.3}$$

and

$$\boldsymbol{A}_{2,mn}=(E_{mn}\mathrm{e}^{-\sqrt{k^2+\chi_2{}^2}\,z}+F_{mn}\,\mathrm{e}^{\sqrt{k^2+\chi_2{}^2}\,z})(\boldsymbol{r}\wedge\mathrm{grad}\ V_{mn}) \quad (0<z) \tag{A2.4}$$

where χ_1 and χ_2 have been defined in (3.231).

We then obtain the magnetic field ocmponents as

$$\boldsymbol{H}=\begin{cases}(-C_{mn}e^{-\sqrt{k^2+\chi_1^2}\,z}+D_{mn}e^{\sqrt{k^2+\chi_1^2}\,z})\sqrt{k^2+\chi_1^2}\,\dfrac{\partial V_{mn}}{\partial x} & \\ (-C_{mn}e^{-\sqrt{k^2+\chi_1^2}\,z}+D_{mn}e^{\sqrt{k^2+\chi_1^2}\,z})\sqrt{k^2+\chi_1^2}\,\dfrac{\partial V_{mn}}{\partial y} & (-L<z<0) \\ (C_{mn}e^{-\sqrt{k^2+\chi_1^2}\,z}+D_{mn}e^{\sqrt{k^2+\chi_1^2}\,z})\,k^2V_{mn} & \end{cases} \tag{A2.5}$$

and

$$\boldsymbol{H}=\begin{cases}(-E_{mn}e^{-\sqrt{k^2+\chi_2^2}\,z}+F_{mn}e^{\sqrt{k^2+\chi_2^2}\,z})\sqrt{k^2+\chi_2^2}\,\dfrac{\partial V_{mn}}{\partial x} & \\ (-E_{mn}e^{-\sqrt{k^2+\chi_2^2}\,z}+F_{mn}e^{\sqrt{k^2+\chi_2^2}\,z})\sqrt{k^2+\chi_2^2}\,\dfrac{\partial V_{mn}}{\partial y} & (0<z) \\ (E_{mn}e^{-\sqrt{k^2+\chi_2^2}\,z}+F_{mn}e^{\sqrt{k^2+\chi_2^2}\,z})\,k^2V_{mn}\,. & \end{cases} \tag{A2.6}$$

As the magnetic field components are continuous at $z=-L$ and $z=0$, we obtain

$$\begin{aligned} &e^{-kL}\,i_{mn}=\sqrt{k^2+\chi_1^2}\,(C_{mn}\,e^{\sqrt{k^2+\chi_1^2}\,L}-D_{mn}e^{-\sqrt{k^2+\chi_1^2}\,L}) \\ &-e^{-kL}i_{mn}=k(C_{mn}e^{\sqrt{k^2+\chi_1^2}\,L}+D_{mn}\,e^{-\sqrt{k^2+\chi_1^2}\,L}) \\ &\sqrt{k^2+\chi_1^2}(C_{mn}-D_{mn})=\sqrt{k^2+\chi_2^2}(E_{mn}-F_{mn}) \\ &C_{mn}+D_{mn}=E_{mn}+F_{mn}\,. \end{aligned} \tag{A2.7}$$

Solving (A2.7), we obtain

$$i_{mn}=-4k\sqrt{k^2+\chi_1^2}\sqrt{k^2+\chi_2^2}\,e^{kL}F_{mn}/\Phi_1 \tag{A2.8}$$

in which Φ_1 has already been given in (3.239).

Adopting λ_1 and λ_2 as defined in (3.240), (A2.8) can be rewritten as

$$i_{mn}=-4k\sqrt{1+\lambda_1^2}\sqrt{1+\lambda_2^2}\,e^{kL}\,F_{mn}/\Phi_1' \tag{A2.9}$$

in which Φ_1' coincides with that given in (3.243).

When we assume $\sigma_1=0$, (A2.9) reduces to

$$(i_{mn})_{\sigma_1=0}=-2k\sqrt{1+\lambda_2^2}\,F_{mn}/(\sqrt{1+\lambda_2^2}+1). \tag{A2.10}$$

In that case, we may define the shielding ratio of the overlying slab by

$$S_e = i_{mn}/(i_{mn})_{\sigma_1=0} \tag{A2.11}$$

in which subscript e means that the ratio represents shielding in the region external to the system of conductors.

(A2.11) can be written as

$$S_e = 2\sqrt{1+\lambda_1^2}(\sqrt{1+\lambda_2^2}+1)e^{kL}/\Phi_1'. \tag{A2.12}$$

Making $\sigma_2 \to 0$, (A2.12) reduces to

$$(S_e)_{\sigma_2=0} = 4\sqrt{1+\lambda_1^2}\, e^{kL}/\Phi' \tag{A2.13}$$

where

$$\Phi' = (\sqrt{1+\lambda_1^2}+1)^2 e^{\sqrt{1+\lambda_1^2}\,kL} - (\sqrt{1+\lambda_1^2}-1)^2 e^{-\sqrt{1+\lambda_1^2}\,kL}. \tag{A2.14}$$

These two equations represent the shielding ratio of a conducting plane slab placed in free space. They are of course in agreement with (3.42) and (3.44) on the conditions that $\mu=1$ and $\sigma_1=\sigma$ ($\lambda_1=\lambda$).

Figure A3 shows how modS_e and mod$(S_e)_{\sigma_2=0}$ with $\Omega_1(=\sqrt{4\pi\sigma_1\omega L})$. The latter is the same as the curve for mod $\bar{e}/e$ in Fig. 3.2. $n=1$, $m=0$, $kL=0.1$ and $\sigma_2=1000\ \sigma_1$ are assumed here. The tendency of the overlying slab to be almost transparent for small values of the induction parameter holds good both for modS_e and mod $(S_e)_{\sigma_2=0}$. The decrease in the shielding ratios as Ω_1 increases starts at smaller

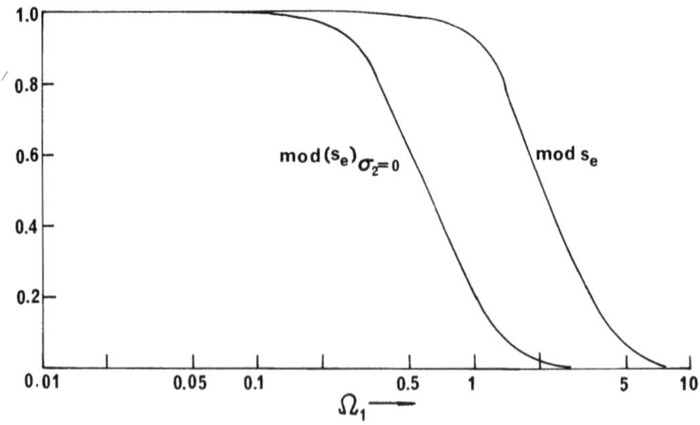

FIG. A3 Changes in modS_e and mod $(S_e)_{\sigma_2=0}$ with Ω_1. $n=1$, $m=0$ and $kL=0.1$ are assumed. $\sigma_2=1000\ \sigma_1$ is also assumed for the first model.

values of Ω_1 for mod $(S_e)_{\sigma_2=0}$ than for mod S_e. As can be seen in the two curves, the oscillating magnetic field is shielded almost entirely by the slab in free space for $\Omega_1 > 2$. However, the same situation for a slab underlain by a semi-infinite medium having a conductivity 10000 times as large as that of the slab requires $\Omega_1 > 10$. It is therefore seen that the underlying conductor weakens the shielding power of a plane slab when the source field arises from inside the underlying conductor.

Corresponding to periods $T=0.1$, 1.0 and 10 years, Ω_1 is calculated as 15.0, 4.77 and 1.50, respectively, for $\sigma_1 = 10\,\mathrm{S/m}\,(=10^{-10}$ e. m. u.) and $L=3000$ km. It is concluded from Fig. A3, therefore, that an earth mantle having the conductivity and thickness cited above and underlain by a core of which the conductivity is $10^5\,\mathrm{S/m}\,(=10^{-6}$ e. m. u.) completely shields the space outside the earth from a geomagnetic secular variation of 1 year period. Even a 100-year period secular variation is reduced by 70 percent in amplitude when it penetrates through the mantle from the core-mantle interface to the earth's surface.

Appendix 3

ELECTROMAGNETIC RESPONSE OF TWO-LAYER SPHERICAL MODEL

Electromagnetic responses as represented R_θ', R_R' and i_1/e_1, which are defined by (3.262) and (3.256) for a spherical two-layer model, are computed for various combinations of τ, q and Ω_1. In the following tables, the computer output for these quantities is listed for $q=0.8$ and 0.4, $\tau=20$. 10, 5 and 0, and Ω_1 ranging from 0.5 to 10 (50 for $\tau=0$).

TABLE A1. $q=0.8$, $\tau=20$ (spherical model).

Ω_1	mod R_0^i	arg R_0^i	mod R_R^i	arg R_R^i	mod i_1/e_1	arg i_1/e_1
0.50	0.1369D+01	4.15°	0.3427D+00	−39.87°	0.1974D+00	16.84°
0.60	0.1389D+01	3.33	0.2895D+00	−41.00	0.2065D+00	13.99
0.70	0.1403D+01	2.66	0.2506D+00	−41.86	0.2133D+00	12.04
0.80	0.1413D+01	2.09	0.2209D+00	−42.55	0.2185D+00	10.64
0.90	0.1421D+01	1.57	0.1974D+00	−43.14	0.2228D+00	9.61
1.00	0.1427D+01	1.10	0.1785D+00	−43.67	0.2262D+00	8.83
1.10	0.1432D+01	0.65	0.1628D+00	−44.15	0.2291D+00	8.23
1.20	0.1436D+01	0.22	0.1496D+00	−44.62	0.2316D+00	7.77
1.30	0.1439D+01	−0.21	0.1384D+00	−45.07	0.2338D+00	7.43
1.40	0.1442D+01	−0.64	0.1287D+00	−45.52	0.2357D+00	7.16
1.50	0.1444D+01	−1.07	0.1203D+00	−45.97	0.2374D+00	6.97
1.60	0.1445D+01	−1.50	0.1129D+00	−46.41	0.2390D+00	6.84
1.70	0.1447D+01	−1.95	0.1064D+00	−46.87	0.2404D+00	6.76
1.80	0.1448D+01	−2.40	0.1005D+00	−47.33	0.2417D+00	6.73
1.90	0.1448D+01	−2.87	0.9528D−01	−47.80	0.2429D+00	6.73
2.00	0.1449D+01	−3.35	0.9054D−01	−48.29	0.2441D+00	6.76
2.10	0.1449D+01	−3.84	0.8624D−01	−48.78	0.2452D+00	6.82
2.20	0.1449D+01	−4.34	0.8231D−01	−49.29	0.2463D+00	6.91
2.30	0.1448D+01	−4.86	0.7872D−01	−49.82	0.2473D+00	7.02
2.40	0.1448D+01	−5.39	0.7541D−01	−50.36	0.2483D+00	7.15
2.50	0.1447D+01	−5.94	0.7236D−01	−50.91	0.2494D+00	7.31
2.60	0.1446D+01	−6.51	0.6954D−01	−51.48	0.2504D+00	7.48
2.70	0.1445D+01	−7.09	0.6691D−01	−52.06	0.2514D+00	7.66
2.80	0.1444D+01	−7.69	0.6447D−01	−52.66	0.2524D+00	7.86
2.90	0.1443D+01	−8.30	0.6218D−01	−53.27	0.2534D+00	8.08
3.00	0.1441D+01	−8.93	0.6004D−01	−53.90	0.2545D+00	8.30
4.00	0.1413D+01	−10.09	0.4417D−01	−61.08	0.2669D+00	10.96
5.00	0.1362D+01	−24.70	0.3405D−01	−69.69	0.2841D+00	13.63
6.00	0.1284D+01	−34.39	0.2675D−01	−79.38	0.3061D+00	15.53
7.00	0.1183D+01	−44.67	0.2112D−01	−89.67	0.3304D+00	16.30
8.00	0.1066D+01	−55.08	0.1666D−01	−100.08	0.3542D+00	16.07
9.00	0.9448D+00	−65.28	0.1312D−01	−110.28	0.3750D+00	15.19
10.00	0.8276D+00	−75.08	0.1035D−01	−120.08	0.3921D+00	13.99

TABLE A2. $q=0.8$, $\tau=10$ (spherical model).

Ω_1	mod $R_0^{\prime}$	arg $R_0^{\prime}$	mod $R_R^{\prime}$	arg $R_R^{\prime}$	mod i_1/e_1	arg i_1/e_1
0.50	0.1240D+01	7.36°	0.6340D+00	−33.26°	0.1507D+00	36.79°
0.60	0.1282D+01	6.24	0.5393D+00	−36.04	0.1669D+00	29.77
0.70	0.1310D+01	5.23	0.4700D+00	−37.76	0.1784D+00	25.13
0.80	0.1331D+01	4.38	0.4171D+00	−39.07	0.1874D+00	21.89
0.90	0.1347D+01	3.62	0.3748D+00	−40.16	0.1947D+00	19.50
1.00	0.1359D+01	2.93	0.3402D+00	−41.09	0.2009D+00	17.67
1.10	0.1368D+01	2.29	0.3113D+00	−41.91	0.2061D+00	16.24
1.20	0.1376D+01	1.69	0.2869D+00	−42.64	0.2106D+00	15.10
1.30	0.1382D+01	1.11	0.2660D+00	−43.32	0.2145D+00	14.18
1.40	0.1387D+01	0.55	0.2479D+00	−43.97	0.2180D+00	13.45
1.50	0.1392D+01	−0.01	0.2320D+00	−44.59	0.2211D+00	12.85
1.60	0.1395D+01	−0.56	0.2180D+00	−45.19	0.2239D+00	12.37
1.70	0.1397D+01	−1.11	0.2056D+00	−45.78	0.2265D+00	11.98
1.80	0.1400D+00	−1.66	0.1944D+00	−46.37	0.2289D+00	11.67
1.90	0.1401D+01	−2.21	0.1844D+00	−46.95	0.2311D+00	11.43
2.00	0.1402D+01	−2.77	0.1753D+00	−47.54	0.2331D+00	11.25
2.10	0.1403D+01	−3.34	0.1671D+00	−48.13	0.2351D+00	11.12
2.20	0.1403D+01	−3.92	0.1595D+00	−48.73	0.2370D+00	11.03
2.30	0.1404D+01	−4.51	0.1526D+00	−49.33	0.2388D+00	10.99
2.40	0.1403D+01	−5.10	0.1462D+00	−49.94	0.2405D+00	10.97
2.50	0.1403D+01	−5.72	0.1403D+00	−50.57	0.2422D+00	10.98
2.60	0.1402D+01	−6.34	0.1348D+00	−51.20	0.2438D+00	11.03
2.70	0.1401D+01	−6.97	0.1297D+00	−51.85	0.2454D+00	11.09
2.80	0.1400D+01	−7.62	0.1250D+00	−52.50	0.2470D+00	11.18
2.90	0.1398D+01	−8.28	0.1205D+00	−53.17	0.2486D+00	11.28
3.00	0.1396D+01	−8.96	0.1164D+00	−53.85	0.2502D+00	11.40
4.00	0.1365D+01	−16.46	0.8532D−01	−61.40	0.2668D+00	13.17
5.00	0.1309D+01	−25.23	0.6547D−01	−70.19	0.2866D+00	15.07
6.00	0.1229D+01	−34.94	0.5120D−01	−79.91	0.3097D+00	16.32
7.00	0.1128D+01	−45.14	0.4028D−01	−90.12	0.3340D+00	16.64
8.00	0.1014D+01	−55.42	0.3170D−01	−100.40	0.3570D+00	16.14
9.00	0.8982D+00	−65.48	0.2495D−01	−110.47	0.3770D+00	15.13
10.00	0.7869D+00	−75.17	0.1967D−01	−120.16	0.3933D+00	13.90

TABLE A3. $q=0.8$, $\tau=5$ (spherical model).

Ω_1	mod R_0'	arg R_0'	mod R_R'	arg R_R'	mod i_1/e_1	arg i_1/e_1
0.50	0.1047D+01	5.53°	0.9391D+00	−15.05°	0.6849D−01	69.38°
0.60	0.1084D+01	6.65	0.8875D+00	−20.30	0.9199D−01	61.84
0.70	0.1125D+01	7.10	0.8230D+00	−25.20	0.1138D+00	54.46
0.80	0.1165D+01	6.96	0.7535D+00	−29.32	0.1323D+00	47.79
0.90	0.1198D+01	6.41	0.6866D+00	−32.54	0.1472D+00	42.12
1.00	0.1225D+01	5.66	0.6267D+00	−34.96	0.1590D+00	37.50
1.10	0.1247D+01	4.85	0.5751D+00	−36.79	0.1685D+00	33.81
1.20	0.1263D+01	4.04	0.5312D+00	−38.23	0.1764D+00	30.88
1.30	0.1276D+01	3.25	0.4937D+00	−39.44	0.1832D+00	28.53
1.40	0.1286D+01	2.48	0.4612D+00	−40.51	0.1891D+00	26.64
1.50	0.1294D+01	1.74	0.4328D+00	−41.50	0.1945D+00	25.08
1.60	0.1301D+01	1.01	0.4077D+00	−42.43	0.1994D+00	23.78
1.70	0.1306D+01	0.30	0.3851D+00	−43.33	0.2039D+00	22.69
1.80	0.1311D+01	−0.41	0.3649D+00	−44.19	0.2080D+00	21.77
1.90	0.1314D+01	−1.11	0.3465D+00	−45.02	0.2119D+00	20.98
2.00	0.1317D+01	−1.81	0.3298D+00	−45.83	0.2155D+00	20.31
2.10	0.1319D+01	−2.51	0.3145D+00	−46.63	0.2190D+00	19.73
2.20	0.1320D+01	−3.21	0.3004D+00	−47.41	0.2222D+00	19.24
2.30	0.1321D+01	−3.92	0.2875D+00	−48.19	0.2253D+00	18.83
2.40	0.1321D+01	−4.63	0.2755D+00	−48.96	0.2283D+00	18.48
2.50	0.1321D+01	−5.34	0.2645D+00	−49.73	0.2311D+00	18.18
2.60	0.1321D+01	−6.07	0.2541D+00	−50.50	0.2338D+00	17.93
2.70	0.1320D+01	−6.80	0.2445D+00	−51.27	0.2365D+00	17.72
2.80	0.1318D+01	−7.54	0.2355D+00	−52.05	0.2391D+00	17.55
2.90	0.1317D+01	−8.29	0.2271D+00	−52.83	0.2416D+00	17.41
3.00	0.1315D+01	−9.04	0.2192D+00	−53.62	0.2441D+00	17.30
4.00	0.1278D+01	−17.15	0.1597D+00	−61.91	0.2679D+00	17.15
5.00	0.1215D+01	−26.19	0.1215D+00	−71.04	0.2918D+00	17.56
6.00	0.1131D+01	−35.92	0.9429D−01	−80.81	0.3164D+00	17.65
7.00	0.1032D+01	−45.96	0.7371D−01	−90.89	0.3403D+00	17.19
8.00	0.9247D+00	−56.01	0.5779D−01	−100.95	0.3620D+00	16.24
9.00	0.8176D+00	−65.83	0.4542D−01	−110.79	0.3804D+00	15.03
10.00	0.7163D+00	−75.33	0.3582D−01	−120.29	0.3954D+00	13.72

TABLE A4. $q=0.8$, $\tau=0$ (spherical model).

Ω_1	mod R_0^l, mod R_R^l	arg R_0^l, arg R_R^l	mod i_1/e_1	arg i_1/e_1
0.50	0.9999E+00	−0.86°	0.5606E−02	89.23°
1.00	0.9985E+00	−3.43	0.2238E−01	86.94
1.50	0.9927E+00	−7.70	0.5006E−01	83.14
2.00	0.9774E+00	−13.56	0.8763E−01	77.93
2.50	0.9474E+00	−20.80	0.1327E+00	71.54
3.00	0.8992E+00	−29.02	0.1815E+00	64.35
3.50	0.8336E+00	−37.72	0.2291E+00	56.87
4.00	0.7559E+00	−46.36	0.2716E+00	49.63
4.50	0.6736E+00	−54.57	0.3068E+00	43.00
5.00	0.5934E+00	−62.14	0.3342E+00	37.20
5.50	0.5195E+00	−69.03	0.3549E+00	32.27
6.00	0.4541E+00	−75.28	0.3703E+00	28.16
7.00	0.3484E+00	−86.28	0.3902E+00	21.93
8.00	0.2709E+00	−95.98	0.4017E+00	17.70
9.00	0.2138E+00	−104.97	0.4090E+00	14.78
10.00	0.1709E+00	−113.63	0.4143E+00	12.74
20.00	0.2431E−01	160.32	0.4493E+00	6.29
30.00	0.4110E−02	76.82	0.4659E+00	4.15
40.00	0.7663E−03	−5.40	0.4742E+00	3.09
50.00	0.1510E−03	−87.16	0.4792E+00	2.47

TABLE A5. q=0.4, τ=20 (spherical model).

Ω_1	mod R_0'	arg R_0'	mod R_R'	arg R_R'	mod i_1/e_1	arg i_1/e_1
0.50	0.1240D+01	6.15°	0.6344D+00	−34.48°	0.2403D−01	49.22°
0.60	0.1283D+01	4.46	0.5397D+00	−37.81	0.2768D−01	46.61
0.70	0.1311D+01	2.81	0.4703D+00	−40.19	0.3115D−01	46.41
0.80	0.1332D+01	1.19	0.4172D+00	−42.26	0.3478D−01	47.45
0.90	0.1346D+01	−0.43	0.3748D+00	−44.21	0.3875D−01	49.09
1.00	0.1358D+01	−2.08	0.3399D+00	−46.10	0.4314D−01	50.94
1.10	0.1366D+01	−3.78	0.3107D+00	−47.98	0.4801D−01	52.77
1.20	0.1371D+01	−5.55	0.2859D+00	−49.88	0.5339D−01	54.46
1.30	0.1375D+01	−7.39	0.2645D+00	−51.82	0.5930D−01	55.91
1.40	0.1376D+01	−9.31	0.2458D+00	−53.82	0.6573D−01	57.10
1.50	0.1375D+01	−11.31	0.2293D+00	−55.89	0.7266D−01	58.01
1.60	0.1373D+01	−13.39	0.2146D+00	−58.03	0.8007D−01	58.64
1.70	0.1369D+01	−15.57	0.2014D+00	−60.24	0.8793D−01	59.01
1.80	0.1363D+01	−17.82	0.1894D+00	−62.53	0.9618D−01	59.14
1.90	0.1356D+01	−20.16	0.1784D+00	−64.90	0.1048D+00	59.05
2.00	0.1346D+01	−22.58	0.1683D+00	−67.34	0.1137D+00	58.77
2.10	0.1335D+01	−25.07	0.1589D+00	−69.86	0.1229D+00	58.32
2.20	0.1322D+01	−27.63	0.1502D+00	−72.44	0.1323D+00	57.71
2.30	0.1307D+01	−30.25	0.1421D+00	−75.08	0.1419D+00	56.96
2.40	0.1290D+01	−32.94	0.1344D+00	−77.78	0.1515D+00	56.11
2.50	0.1272D+01	−35.68	0.1272D+00	−80.53	0.1612D+00	55.15
2.60	0.1252D+01	−38.46	0.1204D+00	−83.32	0.1708D+00	54.12
2.70	0.1230D+01	−41.28	0.1139D+00	−86.15	0.1804D+00	53.02
2.80	0.1207D+01	−44.14	0.1078D+00	−89.02	0.1899D+00	51.86
2.90	0.1183D+01	−47.02	0.1020D+00	−91.91	0.1991D+00	50.67
3.00	0.1158D+01	−49.92	0.9649D−01	−94.81	0.2082D+00	49.45
4.00	0.8738D+00	−78.95	0.5462D−01	−123.89	0.2821D+00	37.50
5.00	0.6147D+00	−106.56	0.3074D−01	−151.52	0.3257D+00	28.59
6.00	0.4217D+00	−132.86	0.1757D−01	−177.84	0.3516D+00	22.84
7.00	0.2863D+00	−158.56	0.1022D−01	156.46	0.3695D+00	19.11
8.00	0.1930D+00	176.00	0.6031D−02	131.02	0.3834D+00	16.51
9.00	0.1293D+00	150.73	0.3592D−02	105.74	0.3948D+00	14.56
10.00	0.8622D−01	125.61	0.2156D−02	80.62	0.4043D+00	13.01

TABLE A6. q=0.4, τ=10 (spherical model).

Ω_1	mod R'_0	arg R'_0	mod R'_R	arg R'_R	mod i_1/e_1	arg i_1/e_1
0.50	0.1048D+01	4.37°	0.9395D+00	−16.21°	0.1582D−01	77.26°
0.60	0.1085D+01	4.96	0.8881D+00	−21.99	0.2161D−01	72.98
0.70	0.1126D+01	4.79	0.8237D+00	−27.51	0.2744D−01	69.18
0.80	0.1166D+01	3.90	0.7542D+00	−32.37	0.3308D−01	66.24
0.90	0.1199D+01	2.51	0.6872D+00	−36.44	0.3857D−01	64.27
1.00	0.1226D+01	0.82	0.6269D+00	−39.81	0.4406D−01	63.12
1.10	0.1246D+01	−1.04	0.5748D+00	−42.68	0.4973D−01	62.55
1.20	0.1260D+01	−2.99	0.5301D+00	−45.26	0.5571D−01	62.32
1.30	0.1271D+01	−5.02	0.4917D+00	−47.71	0.6208D−01	62.25
1.40	0.1278D+01	−7.12	0.4583D+00	−50.11	0.6886D−01	62.20
1.50	0.1282D+01	−9.28	0.4287D+00	−52.53	0.7606D−01	62.10
1.60	0.1284D+01	−11.52	0.4022D+00	−54.97	0.8367D−01	61.92
1.70	0.1283D+01	−13.83	0.3782D+00	−57.46	0.9165D−01	61.62
1.80	0.1280D+01	−16.21	0.3563D+00	−59.99	0.9999D−01	61.21
1.90	0.1275D+01	−18.67	0.3361D+00	−62.58	0.1086D+00	60.67
2.00	0.1268D+01	−21.19	0.3175D+00	−65.21	0.1175D+00	60.02
2.10	0.1259D+01	−23.77	0.3001D+00	−67.89	0.1267D+00	59.26
2.20	0.1248D+01	−26.42	0.2839D+00	−70.62	0.1360D+00	58.41
2.30	0.1235D+01	−29.13	0.2687D+00	−73.40	0.1454D+00	57.47
2.40	0.1220D+01	−31.88	0.2543D+00	−76.22	0.1549D+00	56.45
2.50	0.1203D+01	−34.68	0.2408D+00	−79.07	0.1644D+00	55.36
2.60	0.1185D+01	−37.52	0.2280D+00	−81.96	0.1739D+00	54.22
2.70	0.1165D+01	−40.39	0.2159D+00	−84.87	0.1832D+00	53.04
2.80	0.1144D+01	−43.29	0.2043D+00	−87.81	0.1925D+00	51.82
2.90	0.1121D+01	−46.21	0.1934D+00	−90.76	0.2016D+00	50.57
3.00	0.1097D+01	−49.15	0.1830D+00	−93.73	0.2104D+00	49.31
4.00	0.8302D+00	−78.39	0.1038D+00	−123.16	0.2826D+00	37.33
5.00	0.5853D+00	−106.13	0.5853D−01	−150.98	0.3256D+00	28.52
6.00	0.4020D+00	−132.54	0.3350D−01	−177.44	0.3515D+00	22.83
7.00	0.2731D+00	−158.32	0.1951D−01	156.76	0.3694D+00	19.12
8.00	0.1841D+00	176.19	0.1151D−01	131.25	0.3834D+00	16.51
9.00	0.1234D+00	150.88	0.6854D−02	105.93	0.3948D+00	14.56
10.00	0.8228D−01	125.73	0.4114D−02	80.77	0.4043D+00	13.01

TABLE A7. q=0.4, τ=5 (spherical model).

Ω_1	mod R_0'	arg R_0'	mod R_k'	arg R_k'	mod i_1/e_1	arg i_1/e_1
0.50	0.1003D+01	−0.12°	0.9951D+00	−5.80°	0.1035D−01	87.01°
0.60	0.1006D+01	−0.20	0.9899D+00	−8.32	0.1485D−01	85.72
0.70	0.1010D+01	−0.34	0.9815D+00	−11.26	0.2011D−01	84.22
0.80	0.1017D+01	−0.57	0.9692D+00	−14.57	0.2606D−01	82.55
0.90	0.1026D+01	−0.93	0.9521D+00	−18.19	0.3263D−01	80.75
1.00	0.1037D+01	−1.48	0.9301D+00	−22.06	0.3971D−01	78.85
1.10	0.1050D+01	−2.24	0.9031D+00	−26.10	0.4721D−01	76.92
1.20	0.1064D+01	−3.26	0.8715D+00	−30.21	0.5503D−01	75.00
1.30	0.1078D+01	−4.55	0.8363D+00	−34.33	0.6308D−01	73.14
1.40	0.1092D+01	−6.09	0.7986D+00	−38.39	0.7131D−01	71.36
1.50	0.1104D+01	−7.87	0.7596D+00	−42.34	0.7969D−01	69.68
1.60	0.1113D+01	−9.87	0.7204D+00	−46.14	0.8820D−01	68.10
1.70	0.1121D+01	−12.05	0.6818D+00	−49.81	0.9684D−01	66.61
1.80	0.1125D+01	−14.38	0.6446D+00	−53.33	0.1056D+00	65.20
1.90	0.1127D+01	−16.84	0.6091D+00	−56.73	0.1145D+00	63.83
2.00	0.1125D+01	−19.41	0.5756D+00	−60.04	0.1236D+00	62.50
2.10	0.1122D+01	−22.07	0.5441D+00	−63.26	0.1327D+00	61.18
2.20	0.1115D+01	−24.79	0.5145D+00	−66.44	0.1420D+00	59.86
2.30	0.1106D+01	−27.58	0.4867D+00	−69.57	0.1513D+00	58.55
2.40	0.1095D+01	−30.42	0.4606D+00	−72.69	0.1606D+00	57.22
2.50	0.1082D+01	−33.30	0.4361D+00	−75.79	0.1698D+00	55.89
2.60	0.1067D+01	−36.21	0.4129D+00	−78.89	0.1790D+00	54.55
2.70	0.1050D+01	−39.14	0.3909D+00	−81.99	0.1881D+00	53.20
2.80	0.1032D+01	−42.10	0.3701D+00	−85.09	0.1970D+00	51.85
2.90	0.1012D+01	−45.07	0.3504D+00	−88.19	0.2058D+00	50.50
3.00	0.9917D+00	−48.05	0.3317D+00	−91.29	0.2143D+00	49.16
4.00	0.7542D+00	−77.54	0.1888D+00	−121.57	0.2835D+00	37.04
5.00	0.5338D+00	−105.45	0.1068D+00	−149.83	0.3255D+00	28.41
6.00	0.3675D+00	−132.02	0.6128D−01	−176.60	0.3512D+00	22.81
7.00	0.2499D+00	−157.93	0.3571D−01	157.38	0.3693D+00	19.12
8.00	0.1686D+00	176.49	0.2108D−01	131.72	0.3834D+00	16.52
9.00	0.1130D+00	151.12	0.1256D−01	106.31	0.3948D+00	14.56
10.00	0.7537D−01	125.93	0.7538D−02	81.08	0.4043D+00	13.02

TABLE A8. q=0.4, τ=0 (spherical model).

Ω_1	mod R'_0, mod R'_R	arg R'_0, arg R'_R	mod i_1/e_1	arg i_1/e_1
0.50	0.9997 E +00	−2.01°	0.8240 E −02	88.74°
1.00	0.9950 E +00	−8.00	0.3284 E −01	84.68
1.50	0.9752 E +00	−17.85	0.7252 E −01	78.19
2.00	0.9271 E +00	−31.01	0.1230 E +00	69.70
2.50	0.8441 E +00	−46.47	0.1764 E +00	60.23
3.00	0.7345 E +00	−62.90	0.2242 E +00	51.04
3.50	0.6163 E +00	−79.29	0.2620 E +00	43.05
4.00	0.5051 E +00	−95.17	0.2900 E +00	36.63
4.50	0.4083 E +00	−110.48	0.3107 E +00	31.67
5.00	0.3276 E +00	−125.33	0.3266 E +00	27.87
5.50	0.2617 E +00	−139.86	0.3395 E +00	24.94
6.00	0.2084 E +00	−154.14	0.3505 E +00	22.61
7.00	0.1312 E +00	177.84	0.3687 E +00	19.11
8.00	0.8206 E −01	150.44	0.3832 E +00	16.55
9.00	0.5117 E −01	123.53	0.3948 E +00	14.58
10.00	0.3188 E −01	97.04	0.4043 E +00	13.02
20.00	0.2971 E −03	−157.32	0.4497 E +00	6.29
30.00	0.3108 E −05	−44.72	0.4659 E +00	4.15
40.00	0.3501 E −07	69.92	0.4742 E +00	3.09
50.00	0.4133 E −09	−174.56	0.4792 E +00	2.47

REFERENCES

ASHOUR, A. A., On a transformation of coordinate by inversion and its application to electromagnetic induction in a thin perfectly conducting hemispherical shell, *Proc. London Math. Soc.*, (3) **15**, 557–576, 1965a.

ASHOUR, A. A., The coast-line effect on rapid geomagnetic variations, *Geophys. J.*, **10**, 147–161, 1965b.

BULLARD, E. C. and R. L. PARKER, Electromagnetic induction in the oceans, in *The Sea*, Vol. 4, Ch. 18, John Wiley, New York, 1970.

CHAPMAN, S., The solar and lunar diurnal variation of the earth's magnetism, *Phil. Trans. R. Soc. London, Ser. A.*, **218**, 1–118, 1919.

CHAPMAN, S. and J. BARTELS, *Geomagnetism*, 1049 pp., Oxford Univ. Press, London, 1940.

CHAPMAN, S. and A. T. PRICE, The electric and magnetic state of the interior of the earth as inferred from terrestrial magnetic variations, *Phil. Trans. R. Soc. London, Ser. A*, **229**, 427–460, 1930.

COX, C. S. and J. H. FILLOUX, Two dimensional numerical models of the California electromagnetic coastal anomaly, *J. Geomag. Geoelectr.*, **26**, 257–267, 1974.

DOSS, S. S., and A. A. ASHOUR, Some results on the magnetic field of electric currents induced in a thin hemispherical shell of finite conductivity, with geomagnetic application, *Geophys. J.*, **22**, 385–400, 1971.

FAINBERG, E. B., Electromagnetic induction in the world ocean, *Geophys. Surv.*, **4**, 157–171, 1980.

GUPTA, C. P., V. RAVAL, and J. G. NEGI, Induced current distribution in a conductive two-layer spherical body, *Geophysics*, **38**, 530–544, 1973.

HEWSON-BROWNE, R. C., V. C. L. HUTSON, P. C. KENDALL, and S. R. C. MALIN, New iterative methods of solution of oceanic induction problems, *Phys. Earth Planet. Inter.*, **7**, 431–436, 1973.

HIDE, R. and P. H. ROBERTS, The origin of the main geomagnetic field, in *Physics and Chemistry of the Earth*, Vol. 4, edited by L. H. Ahrens, F. Press, K. Rankama, and S. K. Runcorn, pp. 27–98, Pergamon, London, 1961.

HOBBS, B. A., The calculation of geophysicsl induction effects using surface integrals, *Geophys. J.*, **25**, 481– 509, 1971.

HOBBS, B. A., Electromagnetic induction in the oceans, *Geophys. J.*, **42**, 307–313, 1975.

HOBBS, B. A. and A. T. PRICE, Surface integral formulae for geomagnetic studies, *Geophys. J.*, **20**, 49–63, 1970.

HOBBS, B. A. and A. M. M., BRIGNALL, A method for solving general problems of electromagnetic induction in the oceans, *Geophys. J.*, **42**, 307–313, 1976.

HUTSON, V. C. L., P. C. KENDALL, and S. R. C. MALIN, Computation of the solution of geomagnetic induction problems: a general method, with applications, *Geophys. J.*, **28**, 489–498, 1972.

HUTSON, V. C. L., P. C. KENDALL, and S. R. C. MALIN, The modelling of oceans by spherical caps, *Geophys. J.*, **33**, 377–387, 1973.

KENDALL, P. C., Oceanic induction and shifting the spectrum, *Geophys. J.*, **52**, 201–204, 1978.

KERTZ, W., Leitfähiger Zylinder in transversalen magnetischen Wechselfeld, *Beitr. Geophysik*, **69**, 4–28, 1960.

LAMB, H., *Hydrodynamics*, 6ed. 738 pp., Cambridge Univ. Press, London, 1932.

NEGI, J. G., Electromagnetic screening due to a disseminated spherical zone over a conducting sphere, *Geophysics*, **32**, 69–87, 1967.

NEGI, J. G. and C. P. GUPTA, Models in applied geo-electromagnetics, *Earth-Sci. Rev.*, **4**, 219–241, 1968.

NEGI, J. G. and V. RAVAL, Negative electromagnetic screening by a cylindrical conducting cover, *Geophysics*, **34**, 944–957, 1969.

NEGI, J. G. and S. K. VERMA, Time domain electromagnetic response of a shielded conductor, *Geophys. Prospecting*, **20**, 901–909, 1972.

PRICE, A. T., The induction of electric currents in non-uniform thin sheets and shells, *Quart. J. Mech. Appl. Math.*, **3**, 385–410, 1949.

RIKITAKE, T., The effect of the ocean on rapid geomagnetic changes, *Geophys. J.*, **5**, 1–15, 1961.

RIKITAKE, T., *Electromagnetism and the Earth's Interior*, 308 pp., Elsevier, Amsterdam, 1966.

RIKITAKE, T., Electromagnetic induction within non-uniform plane and spherical sheets, *Bull. Earthq. Res. Inst., Univ. Tokyo*, **45**, 1229–1294, 1967.

RIKITAKE, T., Resonant behaviour of electric currents electromagnetically induced within a nonuniform sheet of conductor, *J. Geophys. Res.*, **73**, 7019–7029, 1968.

RIKITAKE, T., Global electrical conductivity of the earth, *Phys. Earth Planet. Inter.* **7**, 245–250, 1973.

RIKITAKE, T. and I. YOKOYAMA, The anomalous behaviour of geomagnetic variations of short period in Japan and its relation to the subterranean structure. The 6 th report. (The results of further observations and some considerations concerning the influences of the sea on geomagnetic variations), *Bull. Earthq. Res. Inst., Univ. Tokyo*, **33**, 297–331, 1955.

RUNCORN, S. K., The electrical conductivity of the earth's mantle, *Trans. Am. Geophys. Union*, **36**, 191– 198, 1955.

RUNCORN, S. K., On the interpretation of lunar magnetism, *Phys. Earth Planet. Inter.*, **10**, 327–335, 1975.

SOUTHWELL, R. V., *Relaxation Methods in Theoretical Physics*, 248 pp., Oxford Univ. Press., London, 1946.

SOUTHWELL, R. V., *Relaxaiton Methods in Theoretical Physics*, Vol. II, 522 pp., Oxford Univ. Press, London, 1956.

SUGIURA, M. and T. NAGATA, The shielding effect of the ionosphere, *Geophys. Notes, Univ. Tokyo*, **2** (2), 1–9, 1949.

AUTHOR INDEX

SUBJECT INDEX

www.ingramcontent.com/pod-product-compliance
Ingram Content Group UK Ltd.
Pitfield, Milton Keynes, MK11 3LW, UK
UKHW020125290926
15471UKWH00003B/299
* 9 7 8 9 0 4 8 1 8 4 3 3 0 *